Electromagnetic Compatibility in Power Electronics

Electromagnetic Compatibility in Power Electronics

François Costa
Cyrille Gautier
Eric Labouré
Bertrand Revol

Series Editor
Bernard Multon

iSTE

WILEY

First published 2014 in Great Britain and the United States by ISTE Ltd and John Wiley & Sons, Inc.

ISTE Ltd
27-37 St George's Road
London SW19 4EU
UK

www.iste.co.uk

John Wiley & Sons, Inc.
111 River Street
Hoboken, NJ 07030
USA

www.wiley.com

Library of Congress Control Number: 2013951425

British Library Cataloguing-in-Publication Data
A CIP record for this book is available from the British Library
ISBN 978-1-84821-504-7

MIX
Paper from responsible sources
FSC® C013604

Table of Contents

Chapter 1

Phenomena of Perturbation in Electrical Systems

1.1. Electromagnetic perturbations in energy systems

1.1.1. *Introduction*

Power electronic systems are increasingly being used in every field; initially, they were used in the industrial sector and then used increasingly in transportation, services and housing sectors. The flexibility in the control of electrical energy explains this evolution well.

For the purposes of illustration, we estimate that the electrification of service or control functions in an aircraft offers the following gains[1]:

– 10% on the mass;

– 9% on fuel consumption;

– 13% on thrust from the engines;

– 15% on maintenance costs;

– 10% on the buying price.

---

1 According to SAFRAN company, symposium SPEC 2007.

The field of automobiles is also subject to this evolution: the development of hybrid vehicles over the last 10 years and, more recently, the re-emergence of the fully electric car (while waiting for fuel cells vehicles) are evidence of this. Already, a large number of services have been electrified in thermal engine automobiles because of the flexibility of controls (speed variation) and high yield of the electrical systems: power steering, anti-blocking system (ABS), various pumps, window winders, air conditioning (to come).

The introduction of this technology, as a consequence, must take into consideration its implementation constraints; electromagnetic compatibility (EMC) in particular. Indeed, static converters based on power electronics are important sources of electromagnetic perturbations that can occasionally cause malfunctions in their local or distant electronic environment: avionics, navigation systems, reception antennae, etc. Thus, it is important to understand the origin of these phenomena, their mode of propagation and the effects on their potential "victims" in order to optimize the essential reduction or protection devices necessary to conform to the standards of EMC.

A chain of management of the electrical energy is generally organized according to the diagram in Figure 1.1: a primary electrical source powers the energy conversion system (distributed control), which itself powers one or more passive loads or actuators. The link between these components is achieved through conductors or power cables. The converter can itself be a complex device with different levels of conversion and have auxiliary supplies.

The converters carrying out the *processing of electrical energy* (conditioning, control) are based on the use of power electronics in the same manner as microelectronics and *signal processing*. It is noteworthy to observe that these two fields are based on the switching of semiconductors. In the first case, this involves power components (insulated grid bipolar transistor (IGBT), metal oxide silicon field effect transistor (MOSFET, diodes, etc.)) operating with vertical conduction which, in a switching system, confer a very high efficiency to the static converters where they are used; in the other case, this involves heavily integrated lateral components that enable the increase in speed of information processing. In each case, the high-frequency operation of these systems causes electromagnetic perturbations, the disturbance frequencies of which get closer and closer.

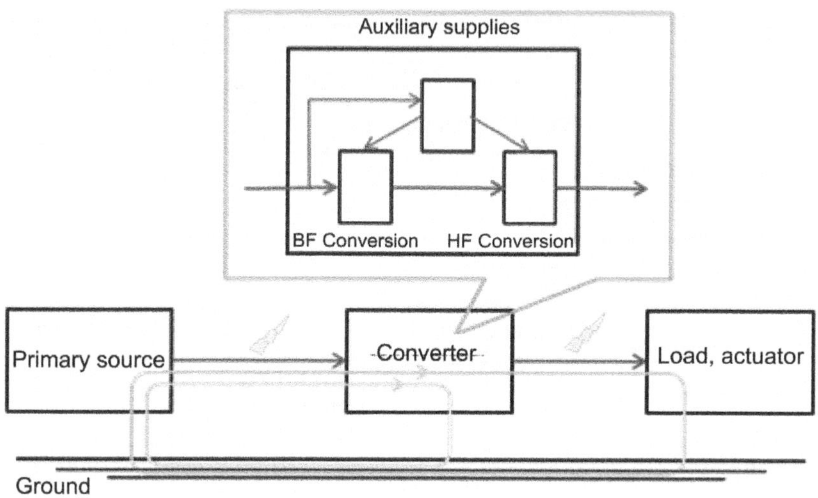

Figure 1.1. *Organization of an power electronics management system*

The consequences of the perturbations emitted by the power devices can be very serious in terms of the reliability and/or security of systems: in an airplane where security depends on electronic localization, communication and flight control systems, the introduction of electrical energy control systems based on power electronics must not threaten the current level of security; a good knowledge of these phenomena is therefore essential in this field.

Near-field or radiative couplings are proportional to the temporal derivative of the electrical quantities: Mdi/dt,

CdV/dt; therefore, as their importance increases, these quantities are naturally bigger and the harmonics of the commuted electrical quantities are of a higher frequency.

Thus, the switching of power semiconductors can cause conducted and emitted electromagnetic perturbations that cover a very large frequency range as shown in Figure 1.2.

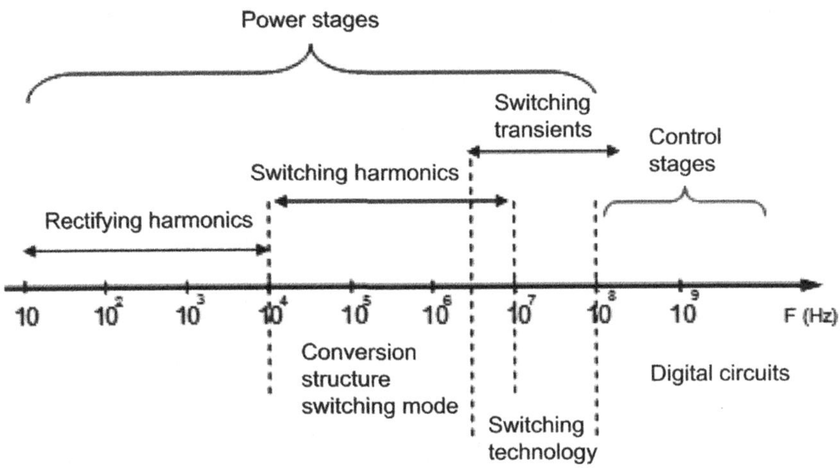

Figure 1.2. *Frequency range of power electronics perturbations*

– At low frequency from the network frequency of 50 Hz: the direct switching of diode or thyristor rectifiers, of triac dimmers, in synchronization with the network frequency, generates perturbations observable up to a few tens of kilohertz. This range is known as the "power grid harmonics".

– At medium frequency (ranging from 10 kHz to 10 MHz): the switching of controlled semiconductors (MOSFET, IGBT) is performed in this range for switched-mode power supplies, choppers and power inverters. The commuted quantities show very quick temporal variations (of the order of a few 10 kV/µs and a few 100 A/µs) with extremely large spectral contents observed over at least four frequency decades: from 10^4 to 10^8 Hz.

– At high frequency: the transients in switching of semiconductors excite the normal modes of very low resistance electrical circuits (necessary for small losses). Thus, very high frequency resonances appear (10 MHz–1 GHz) during each switching between the parasitic inductances of connections (or of magnetic components) and the structural capacitances of the semiconductors.

The reality is more complex than this first classification because in an electronic power device, there are generally several stages of conversion operating at different frequencies (rectifier, high frequency (HF) switch-mode for auxiliary power supplies, medium frequency (MF) switch-mode for power, etc.) that interfere or intermodulate. For illustrative purposes, Figure 1.3 shows the spectrum of the current measured at the input of an upstream switch-mode power supply (black curve) and at the input of a downstream rectifier (gray curve). We can clearly see the contribution of the rectifier starting from 50 Hz and the multiple harmonic peaks that it generates until approximately 10 kHz. Beyond that, we observe (black curve) multiple 15 kHz peaks (switch-mode frequency) that are modulated by the operation of the rectifier and are not modulated on the gray line: the effect of modulation is represented by a certain level of noise at the bottom of the switch-mode harmonic peaks (area circled in dotted line).

These observations show that the electromagnetic perturbations caused by the static converters are not only conducted in the electrical networks and in the cables linking the loads, but are also very easily transmitted by direct radiation, taking into account the amplitudes of the currents and voltages that are in play as well as their frequencies (see Figure 1.1).

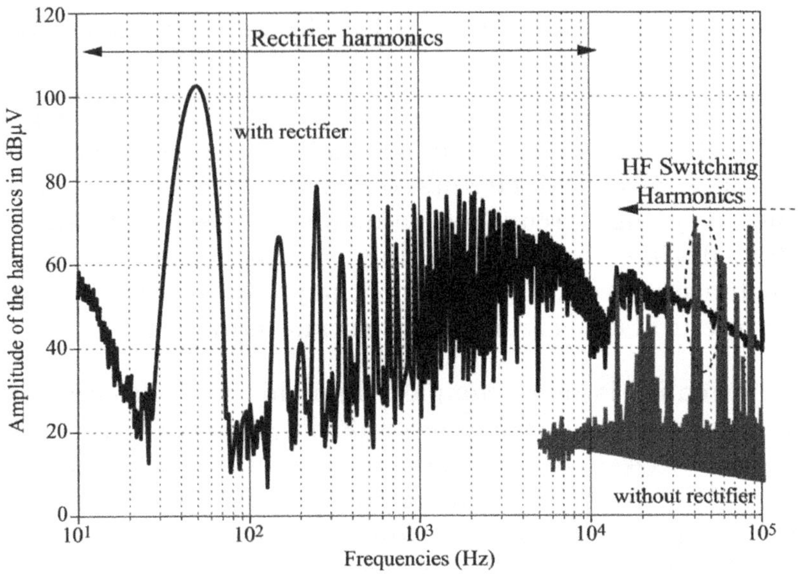

Figure 1.3. *Spectrum of parasite input current of a power supply modulated or not by the input rectifier*

1.2. Power grid harmonics

1.2.1 *Presentation*

In land-based or on-board (aircraft, vessels, etc.) electrical networks, the real shape of the current or voltage wave is never perfectly sinusoidal; real waves include harmonics caused by connected devices and present nonlinear features: diode rectifiers, inductive loads whose magnetic material is saturated over the course of its operation cycle (ballast of fluorescent tubes for example). They therefore summon non-sinusoidal currents which create deformations in the voltage which will, all the while remaining periodic, be deformed by harmonics, generally of odd order.

Figure 1.4 illustrates the propagation mechanism of harmonics in a grid: a nonlinear load creates harmonic currents that, while they travel through the branches of an impedant network, create harmonic voltage drops. The voltage wave is therefore deformed at the observed points. This deformation is evidently bigger as the impedance of the network is also bigger.

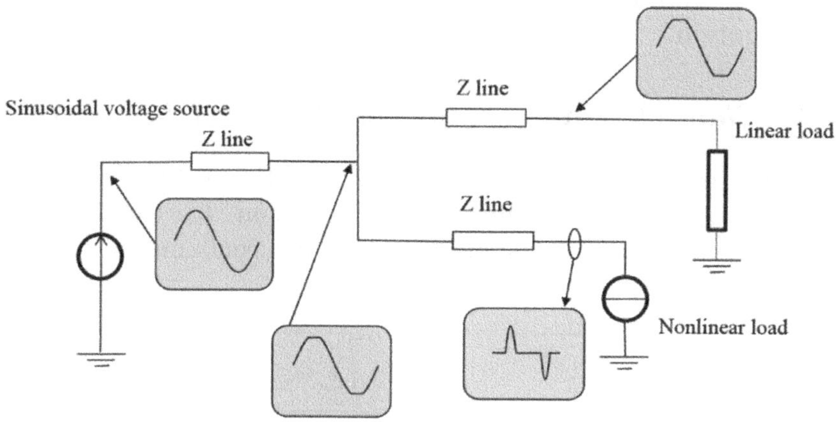

Figure 1.4. *Propagation of harmonics in a network and its consequences on the voltage waveforms*

In addition to the effects resulting from the flow of harmonic currents in the lines of non-zero impedance, the voltage harmonics originate from small imperfections of construction (asymmetries) in the winding of equipment, in other words, rotating machines (motors and alternators) and transformers. These third-order harmonic voltages play a small part, and with low rates, in the origin of the overall harmonic distortions.

For household appliances, it is the accumulation of devices, all in phase and connected to an insufficiently small line impedance, that creates a major harmonic pollution of the network. We can cite, for example, the simultaneous operation of multimedia devices (and of their switch-mode power supply), the constant connection of computers as well as the general use of fluorescent lamps.

The harmonics, being caused by nonlinear loads, are therefore preferentially propagated between phase and neutral on a single-phase network or between the phases of a three-phase network (supposing the load does not have a neutral connection). This is called differential-mode propagation.

1.2.2. *Characterization of the quality of electrical energy*

This pollution is characterized by the *total harmonic distortion* defined either by its relation to the voltage fundamental or by its relation to its root mean square (RMS), as such:

$$\mathrm{TDH_{fund}} = \frac{\sqrt{\sum_{n=2}^{\infty} U_n^2}}{U_1} \qquad \mathrm{TDH_{RMS}} = \frac{\sqrt{\sum_{n=2}^{\infty} U_{n_{eff}}^2}}{U_{eff}} \qquad [1.1]$$

Thus, it is appropriate to be vigilant with the adopted definition when we want to quantify these effects.

Currently in France, the distortion rate, except in certain rare cases, is between:

– 5% and 8% in the low-voltage grid;

– 5% and 7% in the medium-voltage grid;

– 2% and 3% in the high-voltage grid.

The current absorbed by a nonlinear load is defined in the same way by its current distortion rate (we can also find the definition relative to the total RMS value):

$$TDH_{fund} = \frac{\sqrt{I_2^2 + I_3^2 + \ldots}}{I_1} \qquad [1.2]$$

Therefore, we acknowledge that the presence of harmonics contributes to the augmentation of the RMS current, which generally increases losses (joules) in powered loads and in power lines:

$$I = \sqrt{I_1^{\ 2} + I_2^{\ 2} + I_3^{\ 2} + \ldots} = I_1 \sqrt{1 + TDH_{fund}^{\ \ 2}} \qquad [1.3]$$

The quality of the electrical energy is characterized by the *power factor*, defined as the relation between the active power provided to a load and the apparent power, the product of the RMS of voltage and current. In a pure sinusoidal system, ideally the power factor tends toward 1. If we take the case of equipment powered by a purely sinusoidal single-phase voltage and absorbing a non-sinusoidal current, the power factor (per phase) in this scenario is equal to:

$$\lambda = \frac{P}{S} = \frac{V\, I_1 \cos\varphi_1}{V\, \mathrm{I}} = \frac{1}{\sqrt{1 + \mathrm{TDH}_{\mathrm{fund}}^{\ \ 2}}} \cos\varphi_1 \qquad [1.4]$$

where $\cos\varphi_1$ represents the *displacement factor*, in other words the phase shift between the current fundamental and the voltage.

We can therefore see that the presence of harmonics increases the apparent power of a certain appliance or piece of equipment. This influence can equally be rendered as a power balance showing the apparent power, active power, reactive power and deforming power (showing the harmonic distortion). This relation for a grid with q phases is of the form:

$$S^2 = P^2 + Q^2 + D^2 \qquad [1.5]$$

with:

Simple voltage: V

RMS current per phase: I

Current fundamental: I_1

Apparent power: $S = q\ \mathrm{V}\ \mathrm{I}$

Active power: $P = q\ \mathrm{V}\ \mathrm{I}_1\ \cos\varphi_1$

Reactive power: $Q = q\ \mathrm{V}\ \mathrm{I}_1\ \sin\varphi_1$

Deforming power: $D = q\ \mathrm{V}\ \sqrt{\mathrm{I}_2^{\ 2} + \mathrm{I}_3^{\ 2} + \ldots} = q\ \mathrm{V}\ \mathrm{I}_1\ \mathrm{TDH}_{\mathrm{fund}}$

Thus, the presence of harmonics degrades the quality of the electrical energy; it leads to additional losses, vibrations in rotating machines that contribute to the decrease in their longevity.

1.2.3. *Relevant standards for harmonic emissions*

The standard EN 61000-2-2 defines the levels of compatibility for low-voltage public networks and the standard 61000-2-4 defines them for medium- and high-voltage industrial applications. These levels are defined with regard to the network voltage deformations common to different kinds of equipment. We acknowledge that the number of harmonic sources will increase and that the proportion of purely resistive loads (electrical heating), which have a damping effect, will decrease relative to the total consumption. The levels of compatibility for individual harmonics in low-voltage grids are given in Table 1.1.

When several harmonics appear simultaneously, we can express their combined effect by the total harmonic distortion rate (THD_{fund}). Taking into account the levels given in Table 1.1 and that the individual harmonics do not simultaneously reach their level of compatibility, the compatibility level for the global harmonic distortion rate must be limited to 0.08.

Odd harmonics not including multiples of 3		Odd harmonics that are multiples of 3		Even harmonics	
Harmonic order (*n*)	Harmonic voltage (%)	Harmonic order (*n*)	Harmonic voltage (%)	Harmonic order (*n*)	Harmonic voltage (%)
5	6	3	5	2	2
7	5	9	1.5	4	1
11	3.5	15	0.3	6	0.5
13	3	21	0.2	8	0.5
17	2	>21	0.2	10	0.5
19	1.5			12	0.2
23	1.5			>12	0.2
25	1.5				
>25	$0.2 + 0.5 \times 25/n$				

Table 1.1. *Compatibility levels for individual harmonic voltages in low-voltage grids*

As we saw previously, the majority of low-frequency perturbations generated and endured by low-voltage equipment are linked to harmonic perturbations. These perturbations are therefore subject to international and European regulations. We will provide the content of the European regulations regarding the limitations of harmonic currents injected into the public power supply network (current used by appliances ≤ 16 A per phase). This standard is referred to as EN 61000-3-2.

1.2.4. *Classification of appliances*

Concerning the limitation of the harmonic voltage, appliances are classified in the following way:

Class A: balanced three-phase appliances and all other appliances with the exception of those that are labeled as being in one of the following classes.

Class B: portable tools (short-term use).

Class C: lighting appliances including light dimming devices.

Class D: appliances with an input current such that its wave shape at each normalized half-period relative to its peak value is within the range defined in Figure 1.5 (in fact at least 95% of the time) and with an input power of less than 600 W.

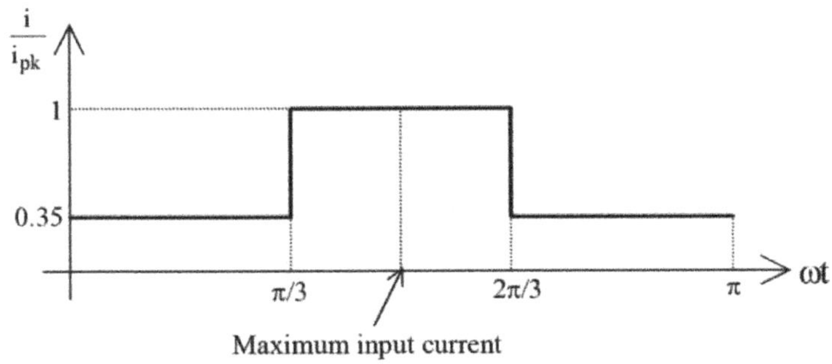

Figure 1.5. *Shape of the current for class D of the EN 61000-3-2 standard*

Regardless of the shape of the input wave, the appliances in class B and class C and motorized phase control appliances are not considered class D appliances.

1.2.5. *The limits of harmonic currents*

They are given in RMS.

Class A appliances: the input current harmonics must not exceed the values indicated in Table 1.2.

Class B appliances: the input current harmonics must not exceed the values indicated in Table 1.2, multiplied by a factor of 1.5.

Class C equipment: we must differentiate lighting appliances and light dimmers. The limits of harmonic currents in lighting appliances must not exceed the limits indicated in Table 1.2.

Harmonic order	Maximum authorized harmonic order (A)
Odd harmonics	
3	2.30
5	1.14
7	0.77
9	0.40
11	0.33
13	0.21
$15 \leq n \leq 39$	0.15×15/n
Even harmonics	
2	1.08
4	0.43
6	0.30
$8 \leq n \leq 40$	0.23×8/n

Table 1.2. *Limits for class A appliances*

For light dimmers that are either independent or incorporated into lamps, the following conditions must apply:

– For independent light dimmers, the values of the harmonic currents incorporated into incandescent lamps must not exceed the values in Table 1.2. When we use a phase-shift command, the firing angle must not exceed 145°.

– For discharge lamps in maximum charge conditions, the value of the harmonic current must not exceed the values defined in percentages in Table 1.2.

– For all positions of the light dimmer, the value of the harmonic currents must not exceed the limits defined in the case of a maximum charge.

Class D appliances: the limits of the harmonic currents are defined in the assigned load conditions. The input current harmonics must not exceed the limits in Table 1.4.

The limits given in Table 1.3 are valid for all appliances whose active input power is greater than 50 W. There are no limits for appliances with an active input power of less than 50 W.

Harmonic order	Maximum authorized harmonic current expressed as a percentage of the fundamental input current of the lighting (%)
Odd harmonics	
3	30 λ (λ: power factor of the circuit)
5	10
7	7
9	5
$11 \le n \le 39$	3
Even harmonics	
2	2

Table 1.3. *Limits for class C appliances*

Harmonic order	Maximum authorized harmonic current per watt (mA/W)	Maximum authorized harmonic current (A)
Odd harmonics		
3	3.4	2.30
5	1.9	1.14
7	1.0	0.77
9	0.5	0.40
11	0.35	0.33
$11 \le n \le 39$	$3.85/n$	0.15×15/n

Table 1.4. *Limits for class D appliances*

1.2.6. *Examples of observations of harmonic currents*

Figure 1.6 shows the measurements[2] taken for several household appliances. We note that they have a high harmonic content. We also observe that the voltage waveform is clipped, so not perfectly sinusoidal, showing the consequences of current harmonics and of an insufficiently low impedance.

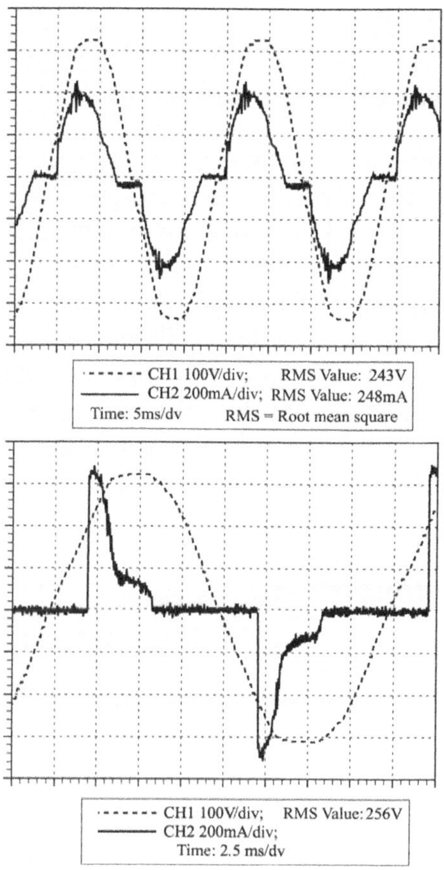

Figure 1.6. *Examples of current and household appliance voltage waveforms measurements highlighting the harmonic content*

2 Measurements provided by J.P. Ferrieux, IUT of Grenoble.

These appliances are produced in very high quantities and therefore contribute to a heavy pollution of distribution networks. We can see especially that energy saving lighting does not represent progress from this point of view; certain office buildings in which the consumption is determined by fluorescent lighting and computing appliances (PC, monitors, printers, etc.) generate a very high level of odd harmonics that require specific reduction methods or the resizing of neutral conductors.

1.2.7. *Fluorescent lighting scenario*

The principle of operation of a fluorescent tube is described in Figure 1.7. The tube has two electrodes at each extremity, and is filled with an inert gas and a small quantity of mercury (in liquid form and in gas form before ignition). The interior of the tube is lined with a mixture of different fluorescent powders. These powders allow the conversion of ultraviolet radiation into several rays situated in visible frequencies. Numerous different powders exist, which can give practically all color temperatures (from hot to cold colors).

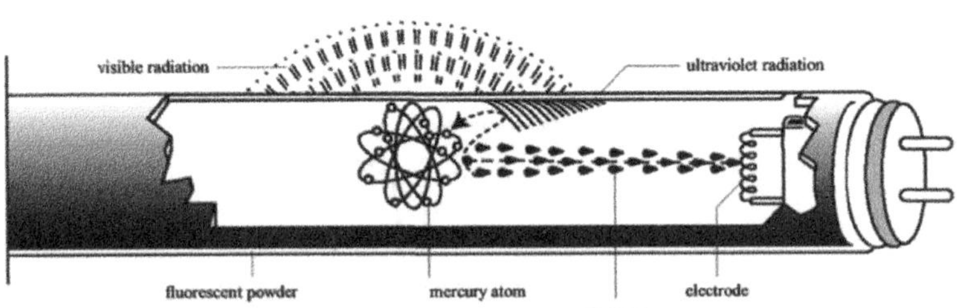

Figure 1.7. *Principle of light emission of a fluorescent tube*

Nevertheless, these lamps have excellent luminous outputs that cannot, as opposed to traditional incandescent lamps, be placed directly on the electrical network. Indeed, the electrodes of the majority of these tubes must be preheated to facilitate ignition (vaporization of mercury). This means that a voltage dedicated to the preheating must be applied before executing the ignition. The ignition of the tube (ignition of the ionization process) is thus realized through the application of a high voltage at the ends of the two electrodes. Moreover, once it is ignited, its negative resistance (Figure 1.8) renders its direct connection to the network impossible. These lamps can therefore only work when combined with an electrical ballast (reactive impedance) allowing for control of the preheating and dimming, and for the limitation of the current circulating in the tube by ensuring stable operation. Indeed, we can observe that due to the distinctive negative resistance, any point of operation where constant voltage is imposed is instable.

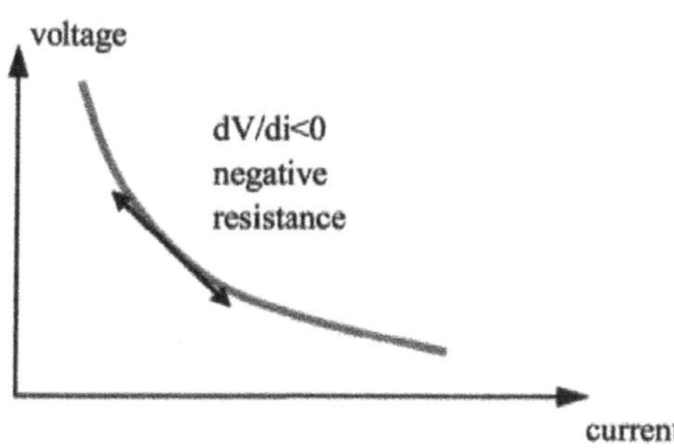

Figure 1.8. *Distinctive negative resistance of a fluorescent tube*

The simplest ballast consists of an iron core inductance operating in tandem with a "starter" for the ignition. The corresponding diagram is shown in Figure 1.9. The combination of the ballast and the lamp acts as a complex form of impedance, but presents a partly positive impedance. This provides a stable operating point in a fixed-voltage grid on the condition that it is placed beyond a voltage limit V_{min}.

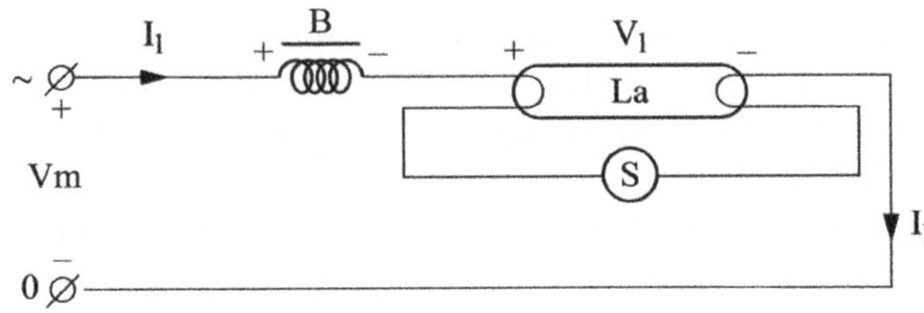

Figure 1.9. *Power supply of the tube through an intermediate ballast*

The curves shown in Figure 1.10 can be determined by only taking into account the fundamental components of the different quantities, particularly for the voltage at the terminals of the tube. The calculations enabling this construction are given below:

$$V_{ballast} = jL\omega I_l \qquad V_{lamp} = f(I_l)$$
$$V = V_{ballast} + V_{lamp} \tag{1.6}$$

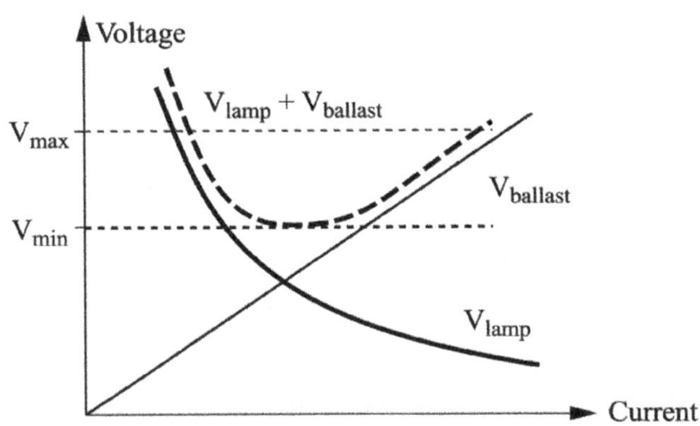

Figure 1.10. *Electrical characteristic of ballast+lamp*

Thus, the ballast can be resized in such a way that a correct operation of the lamp can be ensured in the possible variation range of the grid RMS voltage. Indeed, we must ensure that the minimum voltage is greater than V_{min} and that the maximum voltage V_{max} is compatible with the maximum power that the lamp can consume. This power is in fact limited by the temperature maintained by the electrodes.

When the lamps are powered by the 230 V/50 Hz electric network through a magnetic ballast (as in Figure 1.9), the voltage at the terminals of the lamp and the current that travels through it are not sinusoidal (see Figure 1.11). Each time the current passes through zero, the lamp is neutralized and requires a certain voltage in order to be reignited. The lamp being "hot", this voltage is much weaker than that required for the initial ignition. This re-ignition phase is shown as a slight overvoltage appearing at the start of each half-period. The power factor of the lamp will therefore not exceed 0.8 due to the high deformation of the two quantities.

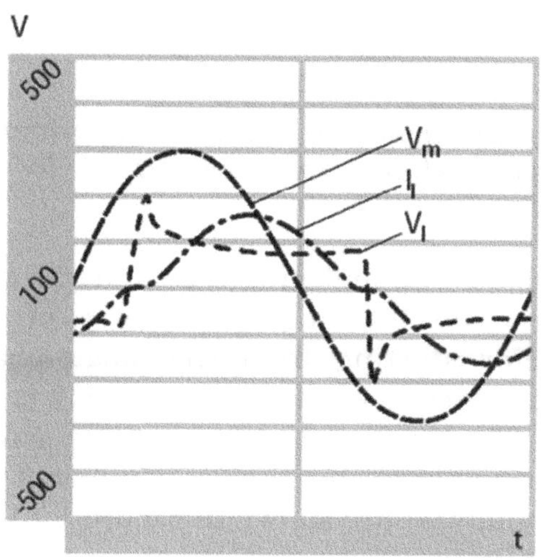

Figure 1.11. *Evolution of the electrical quantities in the fluorescent lamp*

As we can see from the shape of the current as shown in Figure 1.11, the current is not strictly sinusoidal. The deformation of the current is linked both to the square waveform of the voltage at the terminals of the lamp and also to the nonlinearity of the inductance making up the ballast (hysteresis). Measurements performed on tubes in a stabilized state give the following typical values:

Harmonic order	Amplitude (%)	NF EN 61000-3-2
Fundamental	100	100%
3	10	30%
5	3	10%
7	2	7%
9	1	5%

Table 1.5. *Typical values of current harmonic rates for a fluorescent lamp*

Nevertheless, the power factor of the ballast/lamp combination is approximately 0.5. The total third-order harmonic distortion is thus limited to 15%. This equipment therefore conforms to the prescribed values.

1.2.8. *Practical scenario of the improvement of the total harmonic distortion generated by a variable-frequency drive*

Let us consider the following case of a variable-frequency drive using the variable-frequency control of a three-phase induction motor starting from the single-phase electrical network. The principle of this power supply is illustrated in Figure 1.12. The temporal curves of the current and its measured harmonics in this type of equipment with the help of a network analyzer are shown in Figure 1.13.

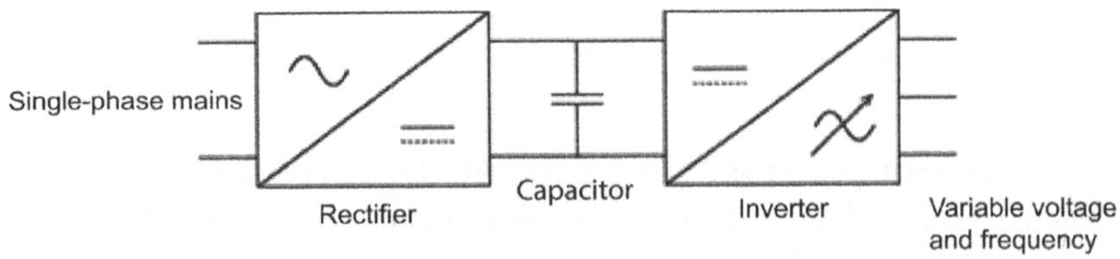

Figure 1.12. *Structure of the variable-frequency drive*

Since this equipment functions on the low-voltage mains and absorbs a current less than 16 A, we can refer to the NF EN 61000-3-2 standard in order to determine the permitted harmonic levels for it. The current absorbed by this equipment is such that it must be considered as a class D appliance. The limits for the permitted harmonic perturbations are given in Table 1.4. The values of these limits from the table can be calculated by using a nominal absorbed power of 750 W and a limit of 2.3 A for the third-order harmonic. It is obvious that in this configuration, the variable-frequency drive does not conform to the regulations. Indeed, if we were to only consider the third-order harmonic, we would obtain a value measured at 2.74 A. The variable-frequency drive therefore does not strictly conform to the regulations. Thus, it is necessary to reduce the harmonic currents.

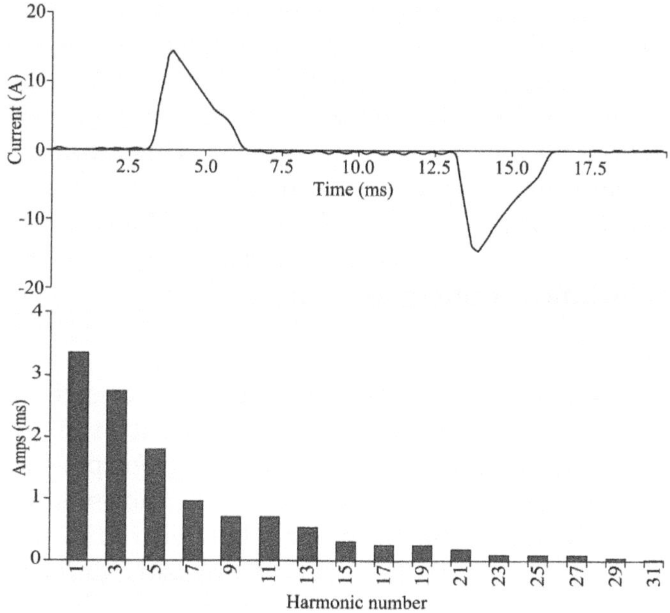

Figure 1.13. *Current absorbed at the input of the variable-frequency drive as a function of time and representation of the harmonics*

One possible solution consists of adding an inductance in series with the variable-frequency drive. The corresponding diagram is given in Figure 1.14.

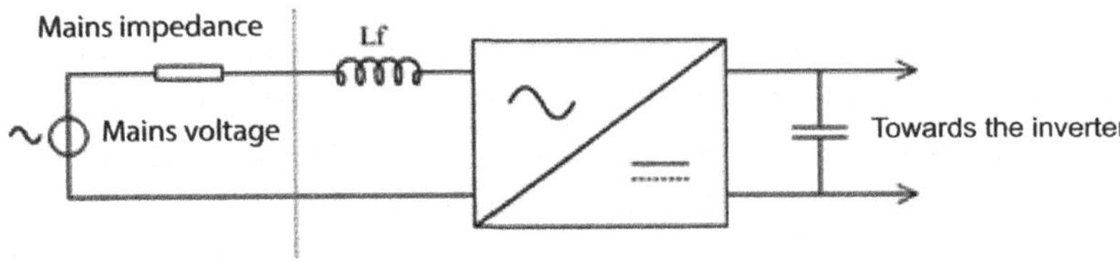

Figure 1.14. *Structure of the variable-frequency drive with an inductance in series*

The presence of an inductance in series helps to reduce the harmonic currents absorbed by the variable-frequency drive. Nevertheless, we note that its value is limited for two reasons: its presence damages the displacement factor of the system through the phase shift of the current fundamental in relation to the voltage. This inductance causes a voltage drop that results in a reduction of the voltage on the continuous bus. For these two reasons, the value of this inductance cannot exceed 10 mH in this example. The results from measurements and the wave shape of the absorbed current are given in Figure 1.15. We can verify that the equipment comprised of the variable-frequency drive/inductance combination now conforms to the EN 61000-3-2 standard.

The analytical calculation of this inductance is quite complex provided we take into account the different constraints we have mentioned (voltage drop, power factor). It is also preferable to resort to the computer simulation. This example has thus been modeled with the equivalent diagram in Figure 1.16 in order to precisely determine the value of L by paying particular attention to the impedance of the source. The values of the simulation components are indicated in the diagram, which help to find the observed electrical values, as shown in Figure 1.17.

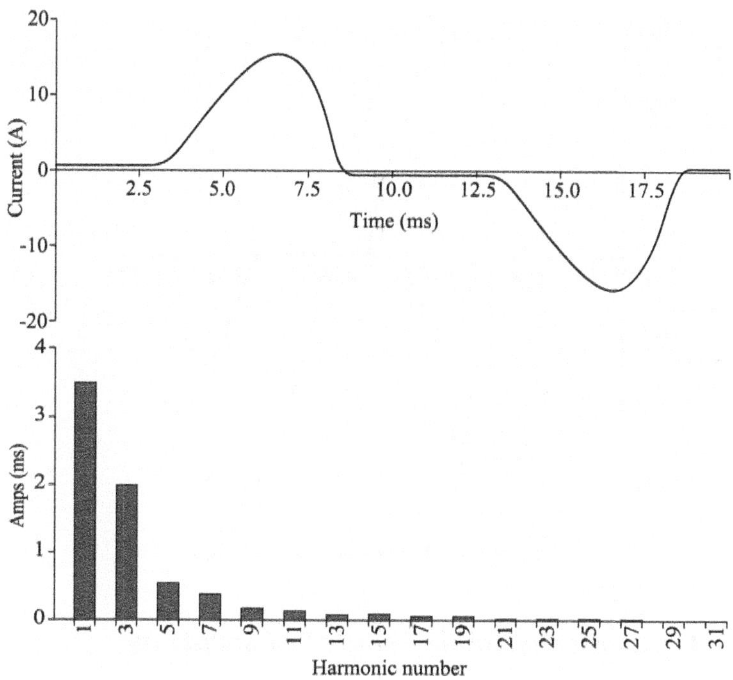

Figure 1.15. *Current absorbed at the input of the variable-frequency drive as a function of time and representation of the harmonics*

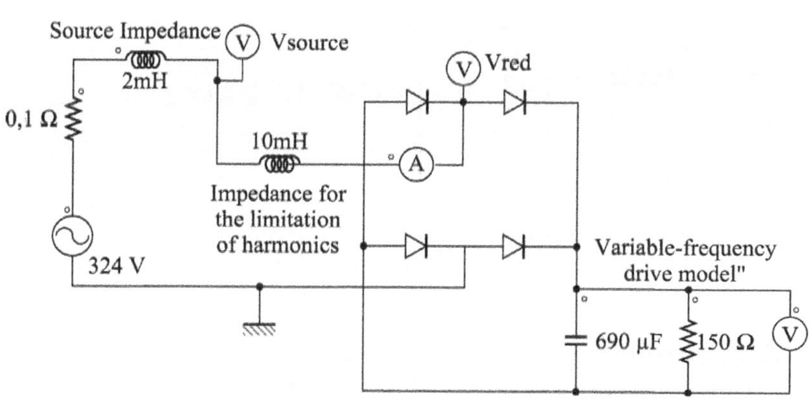

Figure 1.16. *Diagram of the simulation used to determine L*

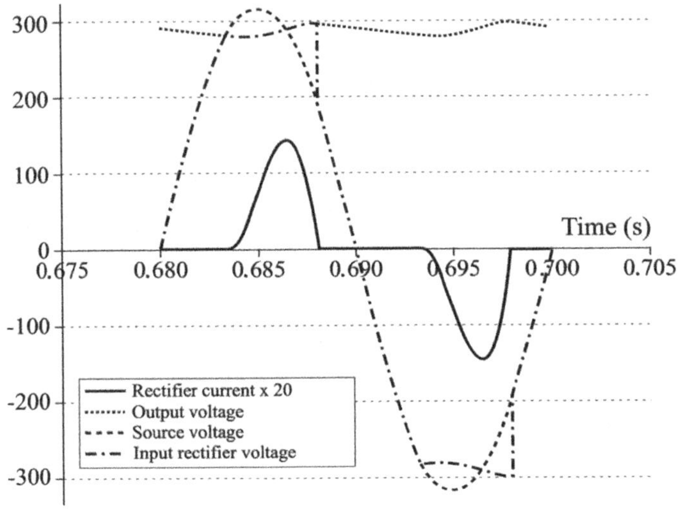

Figure 1.17. *Simulated waveforms*

1.2.9. *Converter with sinusoidal absorption*

The goal is to make sure that the equipment absorbs a current that is as close as possible to a sine curve.

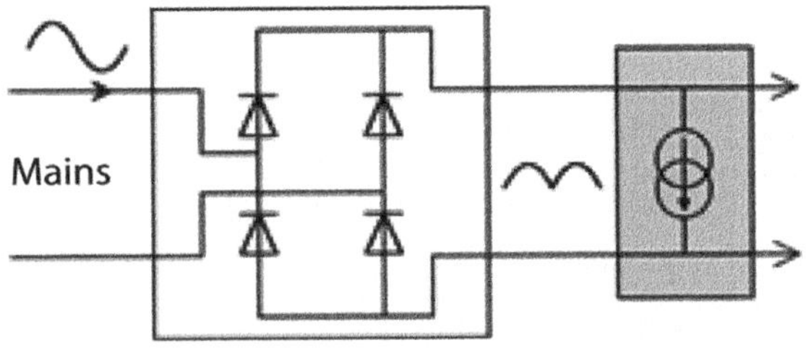

Figure 1.18. *Principle of sinusoidal absorption of current of a converter*

The current absorbed by the equipment, which is placed behind the diode bridge, must therefore be in the shape of a sine curve reflecting the rectified voltage network (see Figure 1.18). If the dynamic of the system is sufficient, the device will absorb a sinusoidal current in phase with the electrical network at all times. The absorbed energy is stored in a capacitor, which itself, provides this energy to the load. The combination of diode bridge/power factor corrector (PFC) behaves, with regard to the mains, as a linear resistive load.

1.2.9.1. *Principle of operation of a PFC*

As we have seen, it involves the creation of a source of current in the shape of a sine curve. The immediate power quantities in play being large, these systems operate on the electrical principle of switching. The most commonly used structures in the field are structures that connect to sources of voltage. The interfacing of the source of voltage and the rectified mains is accomplished because of an inductance. The current control provided by the PFC is therefore executed through the intermediary of a control loop that helps to determine the waveform of the voltage to be generated at each instant. The previously outlined principle is illustrated in Figure 1.19.

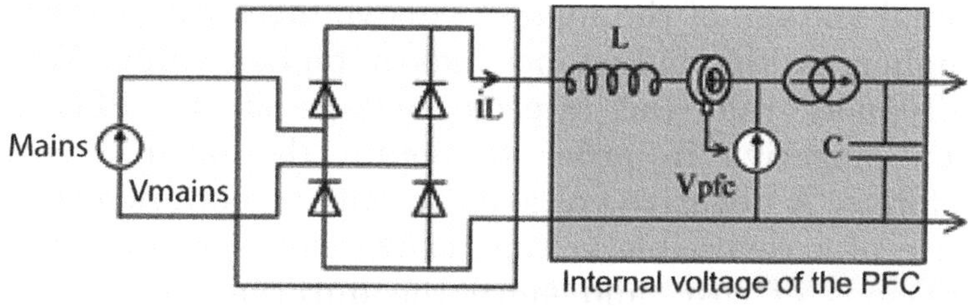

Figure 1.19. *Principle of a power factor corrector (PFC)*

The internal voltage can be calculated at each instant using the following method:

$$v_{PFC} = |v_{main}| - L\frac{di_L}{dt} \quad [1.7]$$

The corresponding waveforms for different inductance values are given in Figure 1.20.

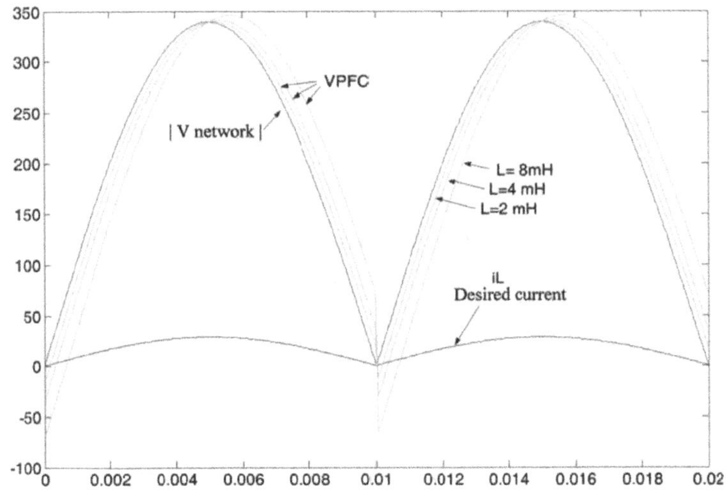

Figure 1.20. *Distinctive waveforms of a PFC*

The observation of these curves shows that the internal voltage of the PFC has a higher peak value than the maximal voltage of the mains. This voltage will incidentally be higher as the inductance has a higher value. We can equally note that the internal voltage of the PFC must become negative in order to obtain the desired current waveforms. A negative voltage can only be obtained through the use of a reversible voltage conversion system, although for reasons of cost, and since the duration for which the voltage is negative is relatively short, the choice consists of a unidirectional voltage structure. The internal voltage of the PFC can therefore not become negative. As a result, this limitation systematically produces a distortion in the waveform of the current as it approaches the crossing through zero.

The structure which enables the generation of the PFC's internal voltage must now be chosen. The immediate power circulating in the PFC is such that the choice of electronic switching is the only solution. The internal source is therefore produced through the intermediary of a chopper via the switching of the voltage at the terminals of the tank capacitor. This corresponds to the diagram in Figure 1.21.

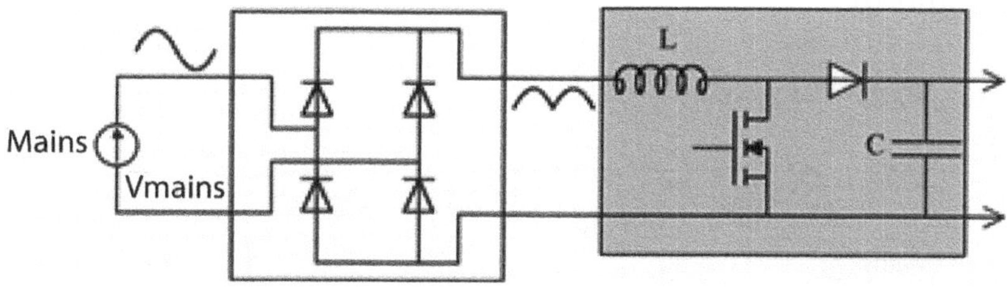

Figure 1.21. *Structure of the BOOST-chopper-derived PFC*

We can note that in these structures, the inductance plays two parts. It acts as a ballast between the rectified mains and the internal voltage of the PFC, as without it the internal voltage would need to be strictly equal to the network voltage and the current would not be able to be regulated. We acknowledge that this inductance also acts as a smoothing inductance for the chopper. Indeed, this system functions on the principle of switching and thus generates square waveforms in the modulated voltage (pulse-width modulation (PWM)). It is obviously not an option to insert high-frequency currents into the mains, which make the smoothing of the currents, as provided by the chopper, necessary. The choice of inductance therefore results in a compromise between the efficiency of the filtering and, as we saw previously, the voltage value at the terminals of the tank capacitor. The control of this system is therefore executed by regulating the current absorbed by the filter. The diagram of the principle behind this regulation is given in Figure 1.22.

We must now turn our attention to the choice of amplitude of the current to be injected into the control loop. We will analyze the system in terms of energy transfer. Supposing that the control loop functions perfectly, the DC-DC converter absorbs a current in the shape of sine curves, the amplitude of which is given by $Iref$. The power absorbed by the system is therefore entirely fixed. Indeed, this power is equal to:

$$P_{abs} = V \, \mathrm{I}_{\mathrm{ref}} / \sqrt{2} \qquad [1.8]$$

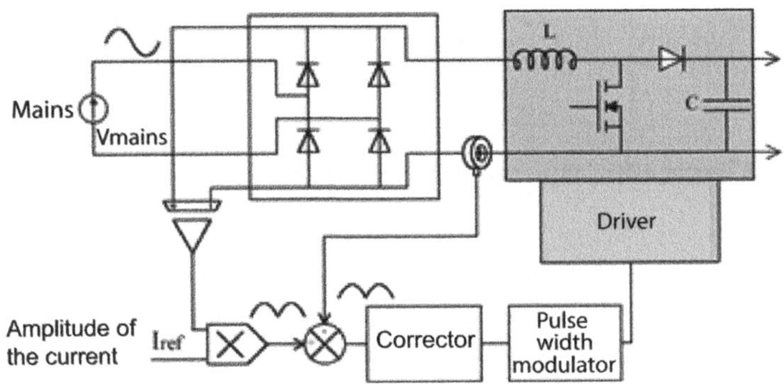

Figure 1.22. *Diagram of current control in a PFC*

The power consumed by the charge depends on the voltage at the terminals of the tank capacitor.

$$P_{load} = f\left(V_c\right)$$

Without control over the input power, these two values cannot be equal. Thus, if $P_{abs} > P_{load}$, the tank capacitor charges itself (the voltage increases). Alternatively, if $P_{abs} < P_{load}$, then the capacitor discharges itself. Therefore, there is no possibility of having stable operation without having control over the amplitude of the absorbed current. Determining the amplitude of the current to be absorbed can be done by using a control loop for the voltage at the terminals of the capacitor. This loop, which has the role of maintaining the capacitor's voltage close to a prescribed value, will modify the amplitude of the absorbed current in order to reach this target. Let us recall that this voltage must be greater than the maximum voltage of the mains, without which the PFC loses control of the shape of the input current. The principle of this control method is shown in Figure 1.23.

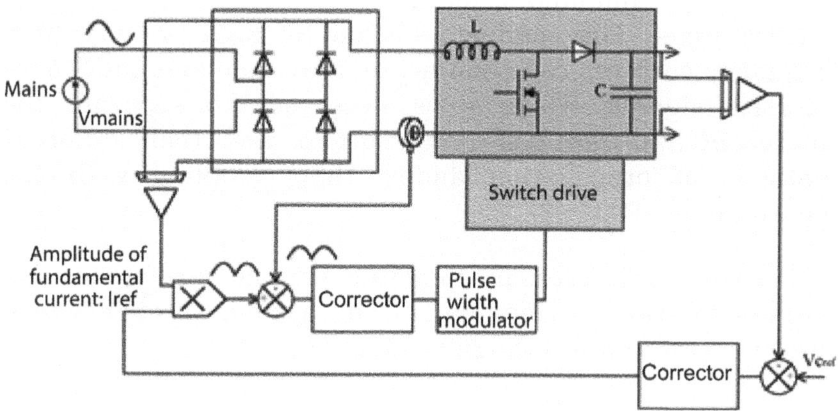

Figure 1.23. *Diagram of the control of the input current and output voltage of a PFC*

This example shows how to approach the problem of harmonic currents in the mains created by a static converter. The implemented solution is based on the use of a chopper which switches at a high frequency compared to the frequency of the electrical network. We will see in Chapter 2 that this principle also generates high-frequency perturbations; a case study will revisit this PFC structure and will analyze it from this aspect, completing this initial low frequency (LF) analysis.

1.3. Common-mode and differential-mode conducted perturbations

If the grid harmonics are in synchronization with the voltage wave of the mains and multiples of 50 Hz, those created by the power-electronic systems are multiples of the switching frequency of each device. Their method of propagation is equally more complex. Indeed, their propagation can be executed on energy lines (like grid harmonics) and also via information transmission lines. Circuit breakers for HF parasite currents can use the earth or ground of appliances, made up of loops with very large surface-areas. The conductors used by high or very high frequency currents can then act as antennae and emit these parasite signals. Thus, it is necessary to separate the *conducted and radiated perturbations* and their different methods of propagation and/or their coupling with the environment [PAU 92].

For this, it is necessary to define a frame of reference and borders for the environment of the system through which they will propagate through conduction.

1.3.1. *Common mode and differential mode*

To study the methods of conducted propagation, it is practical to define a model of electrical links between "boxes" that constitute the borders of the subsystem or equipment in question. We must assign a fixed potential (ground plane, rack, earth) to each "box", to which every subsystem is connected in a more or less local or direct manner. We note that the majority of electrical appliances are equipped with an earth plug; therefore, they are connected to a common reference which is the earth, and the impedance of these links depends on the quality of the earthing. An illustration is given in Figure 1.24 with the example of a switch-mode power supply which powers a load. We can define equivalent perturbation sources (voltage or current) that will allow, via the connection between the converter and the load, the circulation of parasite currents. They loop themselves by following the links in the same direction and by closing themselves through the earth or the common reference, which is the *common-mode* propagation, or in the inverse direction, which is *differential-mode* propagation.

The sources that created these currents are also called common-mode sources or differential-mode sources; in this example, we chose to represent two common-mode voltage sources and a differential-mode source of current. The nature of the source depends on the type of equipment. The parasite currents are determined in part by the impedances through which they are closed off; some are perfectly determined, which in this case would be "physical" differential-mode impedance of the load, while others are more diluted, in which case it would be common-mode impedances that often correspond not to real components but to undesirable coupling effects: capacitive, mutual, insulation faults, etc.

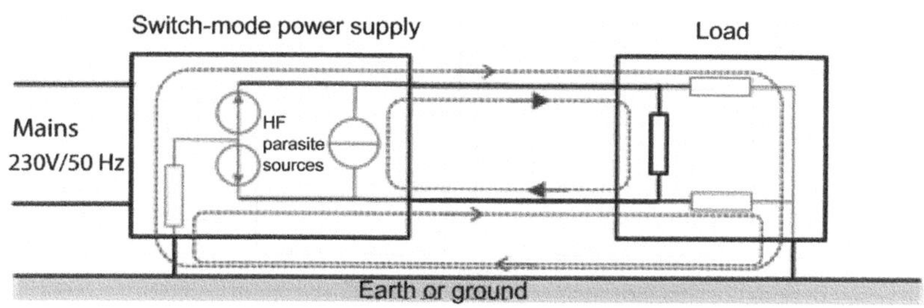

Figure 1.24. *Sources and propagation methods: common mode and differential mode*

This method of representation also allows us to model how perturbations are transmitted through an electrical system, which is known as the coupling method. In a conducted system, in addition to the direct coupling method that is illustrated in Figure 1.24, the main coupling method is said to be through "common impedance".

Coupling through common impedance is possible when the source and the "victims" are connected to the same impedance, for example, devices in derivation on the same supply line or electronic functions routed through the same power supply bus. In differential mode, the perturbing current provokes voltage drops transmitted to the differential-mode "victims", as shown in Figure 1.25. In common mode, the parasite current generated by the source divides itself across the impedances of the source, the common lines and the “victims”, through the earth, also causing undesirable voltage drops. Their levels depend on the relative values of the impedances encountered. Thus, the parasite voltages that appear to be spread out over the common impedance are a result of the set of parasite sources.

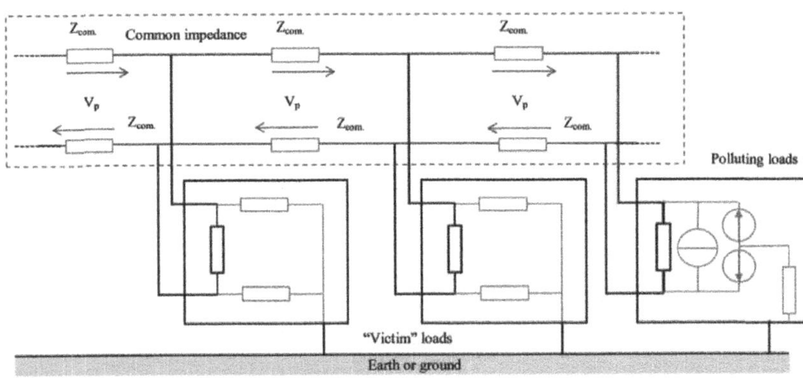

Figure 1.25. *Coupling through common impedance in common mode and differential mode*

In addition to the perturbations transmitted through galvanic contact, there are other methods of perturbation when an electromagnetic field interferes with conductors that connect devices, which is called “field-to-cable” coupling. The two components E and H of the parasitic field contribute to the induction in the cable of the parasite currents or voltages, in common mode and/or differential mode as represented in Figure 1.26. The H component in the field contributes to the appearance:

– of a voltage established by induction in common-mode V_{pMC} created by the magnetic flux traversing the area situated between the cable and the ground plane;

– of a voltage established by induction in differential-mode V_{pMD} due to a flux of H throughout the surface delimited by the two conductors of the link.

These voltages then create common- and differential-mode currents determined by the circuit in question. Generally, the surface of the common-mode loop is much larger than the surface of the differential-mode loop, which results in common mode being more sensitive to this perturbation.

The E component of the field in question contributes to the induction of common-mode currents I_{pMC} and differential-mode currents I_{pMD}.

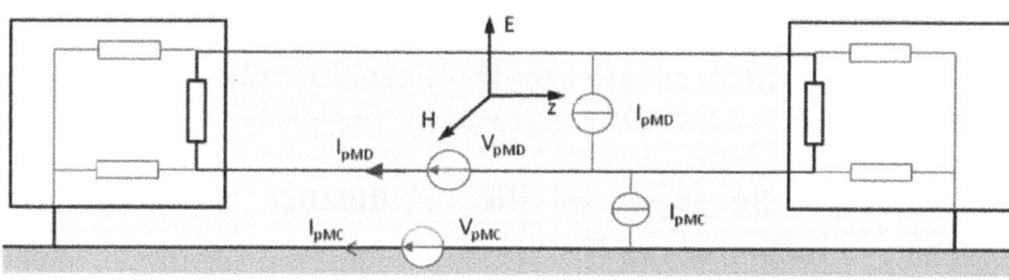

Figure 1.26. *Field-to-cable coupling in common mode and differential mode*

Example number 1: Disruption of measurements in a power-electronics environment.

A voltage probe of an oscilloscope is connected to a switch-mode power supply with an earth link (security of personnel) to carry out a measurement. The oscilloscope is itself earthed. Therefore, it constitutes a loop via the earth as shown in Figure 1.27. The transformer of the switch-mode power supply generates a flux in leaks and the amplitude of which depends on its power and technological make-up. The equivalent diagram represents the effects on the measurement:

– The leakage flux of the transformer establishes a voltage V_{mc} by induction through the loop made up of earth connections and the shielding of the wire:

$$V_{mc} = \int_{earth\_loop} \mu_0 \frac{dH_p}{dt} ds \qquad [1.9]$$

This voltage becomes larger as the frequency of H_p and the surface of the loop become larger.

– The voltage V_{mc} creates a current I_{mc} determined through the impedances of the loop: shielding of the probe, connection between earth and internal connection of the switch-mode power supply.

– The current is created by the parasite voltage V_p that is made up of the value V_0 measured by the probe.

– Finally, the input of the oscilloscope measures the voltage V_m tarnished by the error V_p:

$$V_m = V_0 + \frac{Z_{shielding}}{Z_{blindage} + Z_{earth} + Z_{wiring}} V_{mc} \qquad [1.10]$$

Therefore, we note that the error in measurement depends on the various impedances of the loop and the effect from coupling of the parasite-leak field on the common-mode circuit.

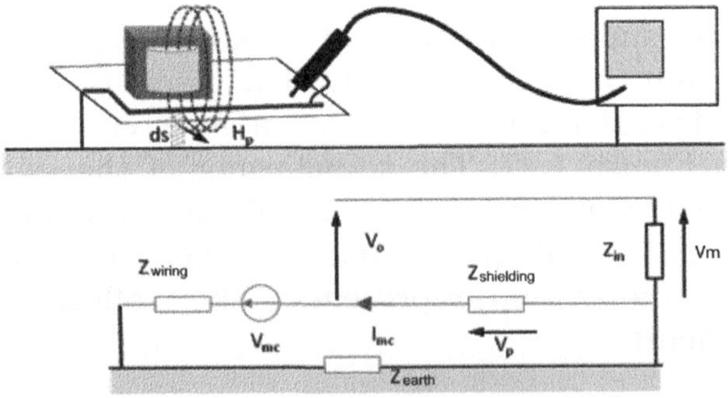

Figure 1.27. *Example of H-field coupling during a measurement*

Example number 2: Impact of a lightning bolt on a household appliance.

The lightning current, even if it does not strike the building directly can have severe consequences through its radiated and conducted effects. It is distinguished by a series of pulses in a current of various amplitudes, which travel through the ionized channel between the thunder cloud and the ground [UMA 87]. To simplify this analysis, we will suppose that the lightning current is represented by a rectilinear conductor traveled by one (or several) pulse(s) of bioexponential current(s) (see Figure 1.30) whose temporal features are shown in Figure 1.28(a) and defined herein:

$$I_{lightning} = I_0(e^{-\frac{t}{\tau_d}} - e^{-\frac{t}{\tau_m}})U(t) \qquad [1.11]$$

U(t) is the echelon function

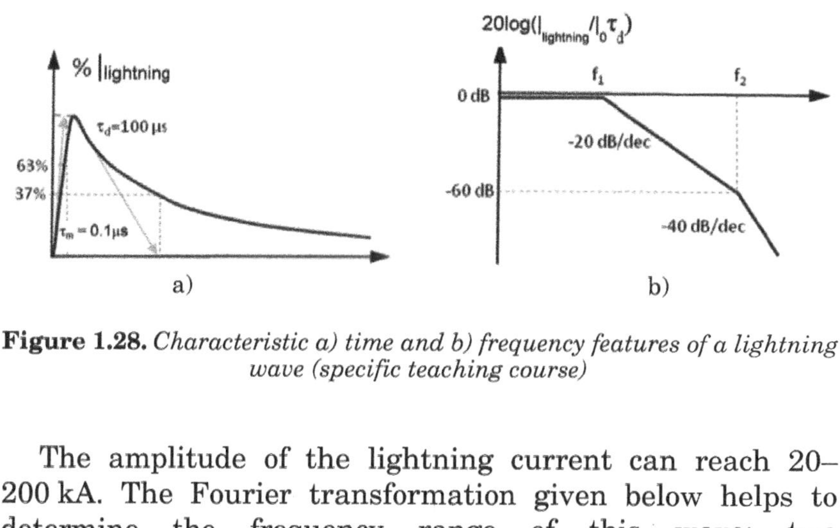

Figure 1.28. *Characteristic a) time and b) frequency features of a lightning wave (specific teaching course)*

The amplitude of the lightning current can reach 20–200 kA. The Fourier transformation given below helps to determine the frequency range of this wave; two distinguishable frequencies f_1 and f_2 appear that determine the frequency ranges over which the spectrum decreases to –20 dB/dec, then –40 dB/dec, as shown in Figure 1.28(b).

$$I_{lightning}(f) = \frac{I_0 \tau_d}{\sqrt{\left[1+(2\pi\tau_m f)^2\right]\left[1+(2\pi\tau_d f)^2\right]}} \quad [1.12]$$

$$f_1 = \frac{1}{2\pi\tau_d} \simeq 1.6\,\text{kHz} \qquad f_2 = \frac{1}{2\pi\tau_m} \simeq 1.6\,\text{MHz}$$

The lightning wave is therefore perturbing for frequencies less than a few megahertz. This representation in the frequency domain also helps to use the expressions of radiation of the antenna obtained in a sinusoidal system to calculate the field that was radiated by a lightning strike.

1.3.1.1. *Emitted effects*

The radiation formula of the Hertz dipole in spherical coordinates at a point P in space is given by the following expression below; we point out in Figure 1.29 the standard orientation of the angles as well as the parameters used.

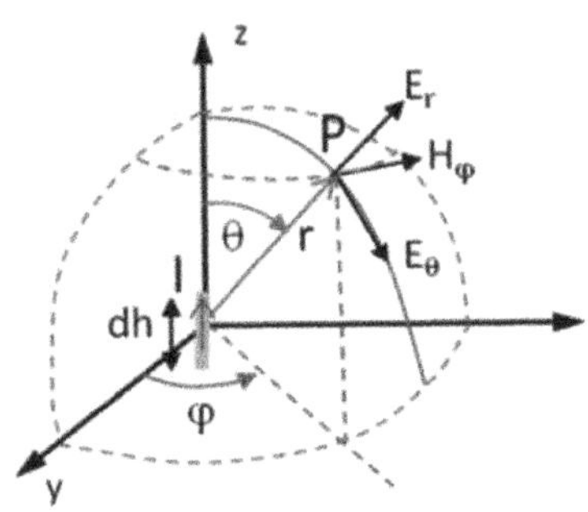

Figure 1.29. *Standard orientation of the Hertz dipole*

$$\begin{cases} E_r = \dfrac{Idh}{4\pi\ j\omega\varepsilon_0}\dfrac{2\cos(\theta)}{r^3}(1+j\gamma r)e^{-j\gamma r} \\ E_\theta = \dfrac{Idh}{4\pi\ j\omega\varepsilon_0}\dfrac{\sin(\theta)}{r^3}\left(1+j\gamma r+(j\gamma r)^2\right)e^{-j\gamma r} \\ H_\varphi = \dfrac{Idh}{4\pi}\dfrac{\sin(\theta)}{r^2}(1+j\gamma r)e^{-j\gamma r} \end{cases}$$

with: $\lambda = \dfrac{c}{2\pi\omega}$

$$\gamma = \frac{2\pi}{\lambda} = \omega\sqrt{\varepsilon_0\mu_0}$$

By simplifying these expressions in the case of a distant field[3], specifically for $\gamma r >> 1$, we obtain the following formulas:

$$E_r \simeq 0$$

$$E_\theta \simeq \sqrt{\frac{\mu_0}{\varepsilon_0}} j\omega\sqrt{\varepsilon_0\mu_0}\frac{I_0 dh}{4\pi}\frac{\sin(\theta)}{r}\cdot e^{-j\gamma r} \qquad [1.13]$$

$$H_\varphi \simeq j\omega\sqrt{\varepsilon_0\mu_0}\frac{I_0 dh}{4\pi}\frac{\sin(\theta)}{r}\cdot e^{-j\gamma r}$$

We represent this situation in Figure 1.30, where the lightning bolt (supposing it is vertical and traveled by a uniform current) of height h, situated at a distance d, radiates an electromagnetic field comparable to a plane wave at the observation point P, center of the loop S_b. We suppose the loop has small dimensions compared to h and d and is found in the field containing the lightning bolt. By using the theory of images in order to take into account the effects of the earth, the components H and E of the field are calculated by integrating the previous formulas over a height of 2 h and by considering that the ratio h/d is such that the distance r remains virtually constant and equal to d:

$$r = \sqrt{d^2 + h^2} \simeq d(1 + \frac{1}{2}\left(\frac{h}{d}\right)^2) \simeq d \;\; error = 5\% \text{ if } {}^{h}\!/_{d} < 0{,}32$$

$$H_\varphi(\omega) = j\omega\sqrt{\varepsilon_0\mu_0}\frac{I(\omega) arctg\left(\dfrac{h}{d}\right)}{2\pi}\cdot e^{-j\gamma d} \qquad [1.14]$$

$$E_\theta(\omega) = \sqrt{\frac{\mu_0}{\varepsilon_0}} H_\varphi$$

3 This hypothesis also signifies that the distance r is large compared to the wavelength, which implies that we consider the frequency components of the current to be sufficiently large: for the numeric examples chosen, they are greater than 100 kHz.

Hence, the following are the time domain formulas:

$$H_{\varphi}(t) = \sqrt{\varepsilon_0 \mu_0} \frac{\frac{dI(t - d\sqrt{\varepsilon_0 \mu_0})}{dt} arctg\left(\frac{h}{d}\right)}{2\pi}$$

$$E_{\theta}(t) = \sqrt{\frac{\mu_0}{\varepsilon_0}} H_{\varphi}(t - d\sqrt{\varepsilon_0 \mu_0}) \qquad [1.15]$$

The term $j\omega$ in the frequency formulas of the H and E fields corresponds to a derivation over time and the term $e^{-j\gamma_0 d}$ corresponds to a delay.

The surface S_b can be made up for example of the loop in the earth conductor of a modem in a home, the shielding of an Ethernet cable of length L_0 situated at a height h_c above the ground and the earth conductor of the computer. The formulas for the resulting current and voltage are given by [UMA 87]:

$$I_{pMC} = C_b h_c \frac{dE}{dt} = C_b \mu_0 h_c \frac{\frac{d^2 I_{lightning}}{dt^2} arctg\left(\frac{h}{d}\right)}{2\pi}$$

$$V_{pMC} = \mu_0 S_b \frac{dH}{dt} = \mu_0 S_b \sqrt{\varepsilon_0 \mu_0} \frac{\frac{d^2 I_{lightning}}{dt^2} arctg\left(\frac{h}{d}\right)}{2\pi} \qquad [1.16]$$

C_b represents the parasite capacitance between the shielding of the Ethernet cable and the earth, the expression of which is given by:

$$C_b = \frac{2\pi\varepsilon_0}{\log(\frac{4h_c}{a})} L_0 \qquad [1.17]$$

where L_0 is the length of the cable, and h_c is its height relative to the ground and its diameter.

Therefore, we note that the second derivative of the lightning current determines these values.

1.3.1.2. *Numerical application*

The lightning current is of the biexponential type, its peak value being 20 kA, $\tau_m = 1$ µs, $\tau_d = 50$ µs. The expression of the maximum value of the derivative is given by:

$$\left(\frac{d^2 I_{lightning}}{dt^2}\right)_{max\,i} \simeq -\frac{I_0}{\tau_m^2} \quad [1.18]$$

The data given for the perturbated circuit consists of: $L_0 = 5$ m, $h_c = 1$ m, $S_b = 5$ m², $C_b = 41.5$ pF, d = 5 km, h = 1.5 km.

We can deduce that the maximum values are $V_{pMC} = 19.4$ V and $I_{pMC} = 48.4$ mA. The waveforms are exponential pulses. These values and their rapid variations are sufficiently significant to excite the normal mode of the line and thus to create common-mode currents in that loop that are susceptible to perturbing the digital connection or even cause irreversible damage to the electronic components.

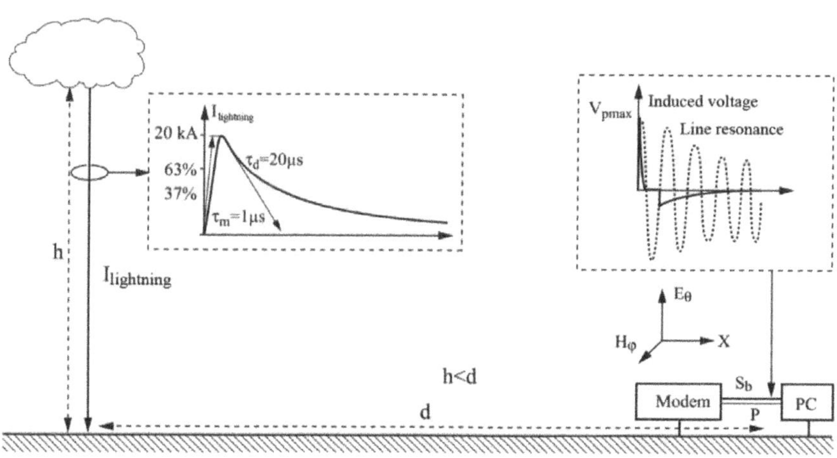

Figure 1.30. *Effects radiated by a lightning strike on a circuit*

1.3.1.3. *Conducted effects*

The ground has a resistance ρ which depends on its nature and humidity rate; therefore, a lightning strike will create a voltage wave V(r) that will develop itself radially from the impact point according to the following law:

$$V(r) = \frac{\rho I}{2\pi r} \quad [1.19]$$

The lines of potential are spread along semi-hemispherical curves as shown in Figure 1.31.

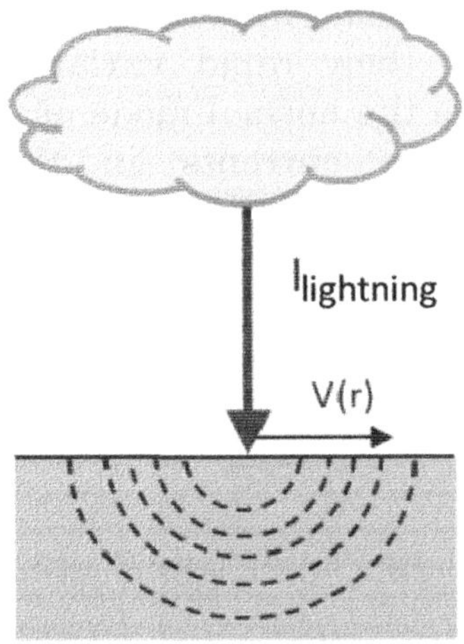

Figure 1.31. *Voltage resulting from a lightning strike on the ground*

Thus, an electric potential is developed between two earthed points and determined by the distance between these two points, the resistance of the ground and the intensity of the lightning. This voltage is responsible for the shocking of animals in fields or the destruction of electrical equipment in buildings where there are multiple distant earth connections.

1.3.2. *Crosstalk*

Crosstalk, sometimes referred to as "cable to cable coupling", shown in Figure 1.32 is mentioned when connections are close together and perturb each other through capacitive coupling (C_{12}) and/or through mutual inductance (L_{12}). These couplings are distributed all along conductors 1 and 2, but if the length L_0 is small compared to the wavelength of the perturbing voltage V_0, we can establish the equivalent lumped-elements circuit as shown in Figure 1.32.

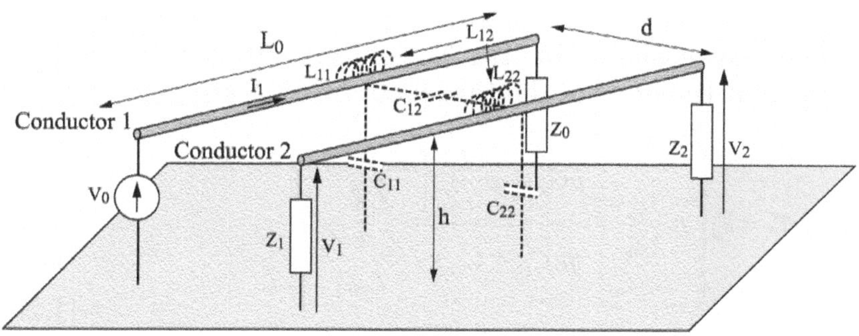

Figure 1.32. *Crosstalk between two conductors*

In this hypothesis, the source of perturbations V_0 injects a capacitive current via C_{12} and an inductive current resulting from mutual inductance L_{12} into the victim circuit comprised of conductor 2 and the impedances Z_1 and Z_2. In Figure 1.33, we represent an electric model of crosstalk coupling where the effects are distributed symmetrically with regard to the middle of the opposite conductors; we suppose that the conductors have a low impedance compared to those of the end loads, which is generally the case. We can see that the effect from the crosstalk is different at the start and at the end of the target connection: the capacitive and inductive parasite currents add to each other in one case and cut themselves off in the other.

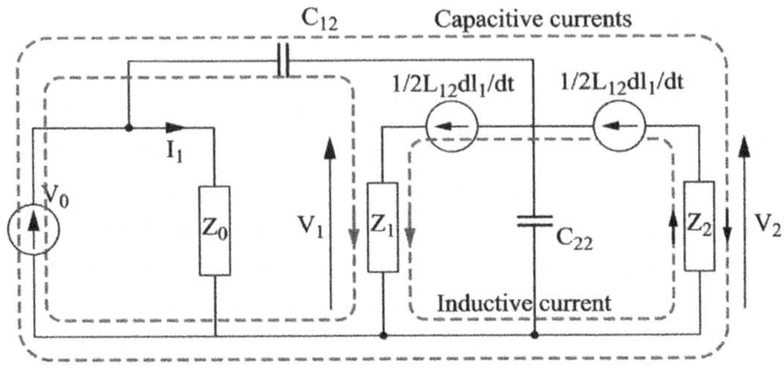

Figure 1.33. *Electrical diagram of crosstalk coupling between two conductors*

The expressions for the parasitic voltages V_1 and V_2 induced at the ends of the target connection are given by:

$$V_1 = V_0 \left(p\frac{L_{12}}{2Z_0} + \frac{pC_{12} + p\frac{L_{12}}{2Z_0}\left(\frac{Z_1 - Z_2}{Z_1 Z_2}\right)}{p(C_{12} + C_{22}) + \frac{Z_1 + Z_2}{Z_1 Z_2}} \right)$$

$$V_2 = V_0 \left(-p\frac{L_{12}}{2Z_0} + \frac{pC_{12} + p\frac{L_{12}}{2Z_0}\left(\frac{Z_1 - Z_2}{Z_1 Z_2}\right)}{p(C_{12} + C_{22}) + \frac{Z_1 + Z_2}{Z_1 Z_2}} \right) \quad [1.20]$$

Example number 3: When the diameter of the conductors is small compared to h and d, the linear expressions l_{ij} in the case of cylindrical conductors of diameter ϕ in the air above a ground plane are given by [GUE 02]:

$$l_{ii} = \frac{\mu_0}{2\pi}\ln\left(\frac{4h}{\phi}\right) \qquad l_{ij} = \frac{\mu_0}{4\pi}\ln\left(1 + \frac{4h^2}{d^2}\right) \quad \text{for } i \# j \quad [1.21]$$

The terms c_{ii} and c_{ij} are obtained through inversion of the matrix $[l_{ij}]$ given by:

$$\left[c_{ij}\right] = \varepsilon_0 \mu_0 \left[l_{ij}\right]^{-1} \quad [1.22]$$

and finally the expressions corresponding to the length L_0:

$$\left[L_{ij}\right]=L_0\left[l_{ij}\right] \qquad \left[C_{ij}\right]=L_0\left[c_{ij}\right] \qquad [1.23]$$

If we suppose that the impedances are identical at the ends of the target connection and that all impedances are resistances, the previous expression is simplified to:

$$\begin{aligned} V_1 &= V_0\left(p\frac{L_{12}}{2R_0}+\frac{p\frac{R}{2}C_{12}}{1+p\frac{R}{2}(C_{12}+C_{22})}\right) \\ V_2 &= V_0\left(-p\frac{L_{12}}{2R_0}+\frac{p\frac{R}{2}C_{12}}{1+p\frac{R}{2}(C_{12}+C_{22})}\right) \end{aligned} \qquad [1.24]$$

Numerical application:

L_0 = 10 cm, h = 10 mm, d = 10 mm, ϕ_1 = 3 mm, ϕ_2 = 1 mm, the source of the disruption is a triangular alternative voltage generator with an amplitude of 25 V and a frequency of 500 kHz, R_0 = 5 Ω, R_1 = R_2 = 50 kΩ.

We deduce that: L_{11} = 52 nH, L_{22} = 74 nH, L_{12} = 32 nH, C_{11} = 3 pF, C_{22} = 2 pF, C_{12} = 1.5 pF.

Figure 1.34 shows a simulation of this crosstalk, where we can clearly see that the voltages at the two ends are in the region of 2 V and different to each other. In these simple hypothetical cases, coupling is essentially due to derived effects; it is therefore increasing with the frequency.

According to the configuration of the influencing conductors and their shape, the amplitude of the currents and the voltage in the source conductor, one coupling mode can be dominant (magnetic or electrostatic); this hypothetical case therefore consists of ignoring either C_{12} or L_{12} in the model.

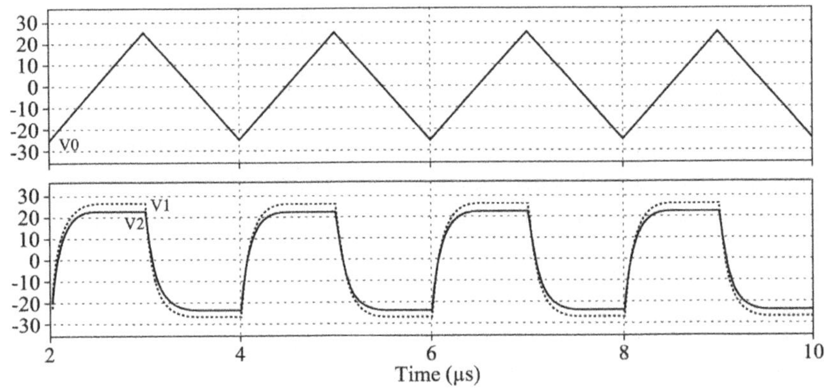

Figure 1.34. *Simulation of crosstalk coupling between two conductors*

1.4. Measuring electromagnetic perturbations

We distinguish two categories of measuring tools used in EMC. The first category involves the measuring of perturbations conducted at high frequencies; this includes the line impedance stabilization network (LISN) and the passive current sensors based on the principle of the current transformer. In the second category, we can find various types of antennae used to measure in the near or distant field. In every case, the signal from the sensor is analyzed in the time domain (oscilloscope) and more frequently in the frequency domain using the heterodyne spectrum analyzer.

1.4.1. *The line impedance stabilization network*

The LISN is connected to a filter fitted between the device being tested and the network supplying the energy. It has a dual purpose. In a standardized frequency range (150 kHz–30 MHz), it must isolate the network, in which common-mode and differential-mode perturbations may exist, from the equipment being tested. Nonetheless, it must provide an input-to-output voltage drop less than 5% of the nominal voltage to the frequency of the network when it is traveled by the nominal current. Finally, it must present a terminating impedance that is constant with regard to the high-frequency perturbations emitted by the test device, in both common and differential modes, independent of the impedance introduced by the energy network. This impedance is standardized by CISPR (*Comité international spécial des perturbations radioélectriques* – Special international committee on radio interference) [INT 03a]. Figure 1.35 defines the limits in variation of this impedance, measured between an output terminal and the earth in the frequency range of 10 kHz–100 MHz. This range is divided into three bands (A, B, C). The corresponding diagram relative to each band is shown in Figure 1.35. The structure of the LISN is represented in Figure 1.36.

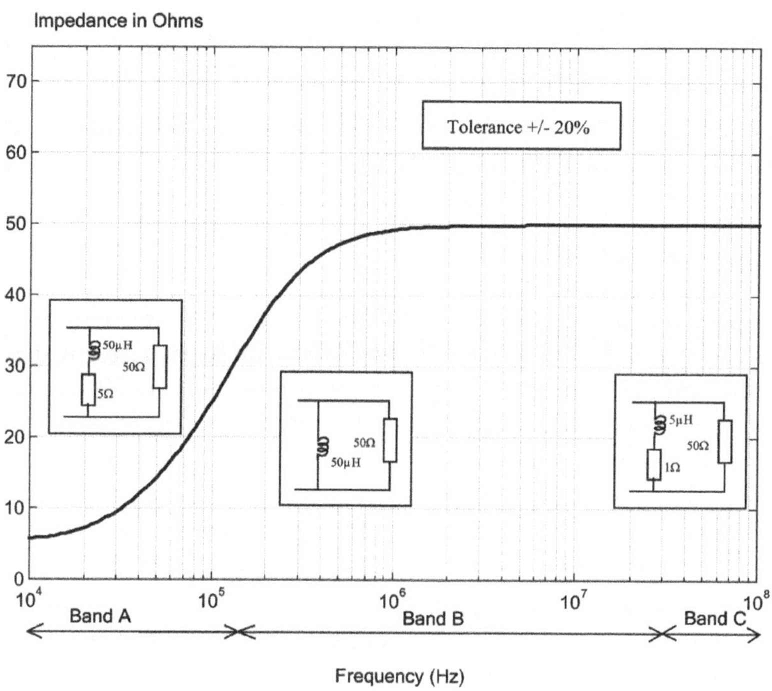

Figure 1.35. *Common-mode impedance of the LISN*

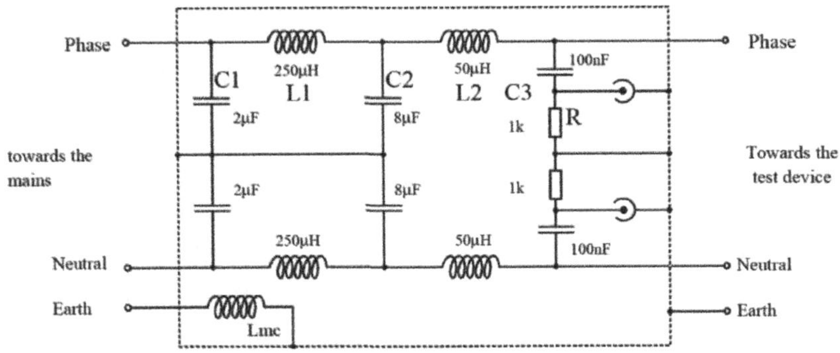

Figure 1.36. *Diagram of the principle of a single-phase LISN (according to document of. Chase Electronics)*

The filter made up of L1, C1 and L2, and C2 is a low-pass filter that eliminates perturbations coming from the mains. The filter C3, R is a high-pass filter that enables the termination of high-frequency perturbations in the resistance R_0 of 50 Ω. The LISN can be used for currents with intensities up to 100 A. Past that, and for band A, it is possible to use a current probe.

1.4.2. *Current sensors*

The current sensors used in metrology in static conversion devices must have a very large bandwidth, in the hundreds of megahertz, must be capable of measuring high currents (up to several kiloamperes) and be insensitive to electromagnetic strikes released by the converter being tested. They are also used for measuring conducted parasitic currents. In this case, they must have a large bandwidth and not be prone to perturbations. Experience shows that passive sensors, based on the principle of the current transformer, are best suited to these metrological constraints, since knowledge of the DC component is not essential. The lumped-element electric model of such a device is shown in Figure 1.37.

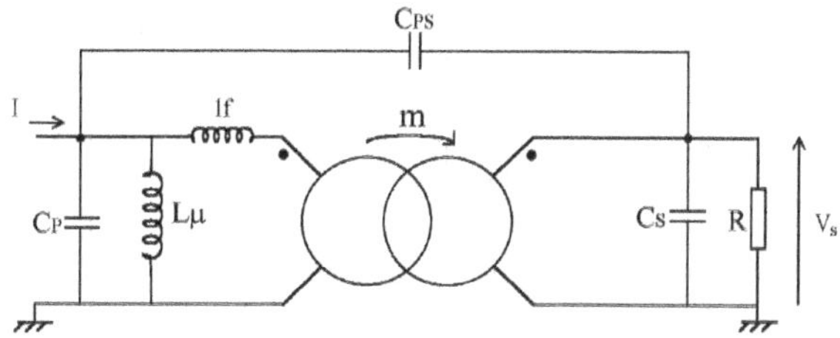

Figure 1.37. *Model of the current transformer*

This model allows us to understand the performance and limitations of this device. They involve the low cutoff frequency F_L, the linearity outside saturation, high-frequency operation and the resistance to electromagnetic perturbations.

1.4.2.1. *Behavior at low frequency and the effects of saturation*

At low frequency, the current transformer has a high-pass behavior. Its voltage transfer function V_s/I and its low cutoff frequency are given by the following expressions:

$$\frac{V_s}{I}=\frac{R}{m}\frac{j\frac{m^2 L_\mu}{R}\omega}{1+j\frac{m^2}{R}\left(L_\mu+lf\right)\omega} \qquad [1.25]$$

$$F_L=\frac{1}{2\pi}\frac{R/m^2}{L_\mu}=\frac{k}{2\pi\mu_0\mu_r n\frac{A_e}{l_e}} \qquad [1.26]$$

where A_e is the section of the magnetic circuit, l_e is the effective length of the magnetic circuit, n is the number of secondary turns, m is the transformation ratio (generally, there is one primary turn, hence $m = n$) and $k = R/n$.

These expressions are valid for a low-frequency band (<a few tens of kilohertz) in which the electrostatic effects

are negligible. The gain of this transfer function is given in Figure 1.38 (BF zone of the curve).

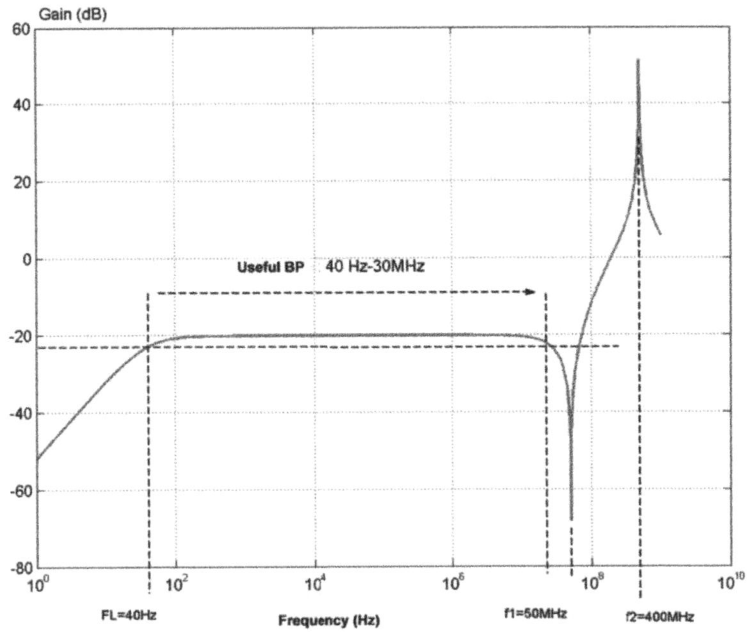

Figure 1.38. *Transfer function of the current transformer obtained through the model*

When the measured current has a DC or low-frequency component (outside its bandwidth), the magnetic circuit can be saturated, which translates into a distortion of the output signal. The expression below defines, in a sinusoidal system with a DC component, the limits in the linear operation of the device as a function of the quantities involved (DC current I_{DC}, maximum AC current I_{ACmax}, frequency ω and the specifications of the magnetic circuit: B_M, μ_r, specific inductance A_L, magnetic length l_e).

$$I_{AC\max i} = \left(\frac{B_M}{\mu_0 \mu_r} l_e - I_{DC} \right) \sqrt{1 + \frac{(n^2 A_L \omega)^2}{R}} \qquad [1.27]$$

In pulsed current, the saturation of the magnetic circuit is characterized by the maximum of the product I.T. Its expression is given by:

$$(\mathrm{I} \cdot \mathrm{T})_{\max} = \frac{\mathrm{n}^2}{\mathrm{R}} \cdot \mathrm{B_M} \cdot \mathrm{A_e} \qquad [1.28]$$

1.4.2.2. *Behavior at high frequency and with regard to perturbations*

A third limitation appears for high-frequency operation. The expressions [1.26] and [1.27] show that to reduce F_L and increase I_{ACmax}, we must increase the number of winding turns n and the section of the magnetic circuit A_e. However, this is contradictory considering the high-frequency performances of the sensor. Indeed, the length of the secondary winding increases, just like the electrostatic phenomena (capacitances of the primary and secondary windings Cp, Cs), which limits the high-pass band of the sensor, as the current transfer function shows (see the HF section of the curve in Figure 1.38):

$$\frac{Vs}{I} = \frac{1 + l_f.C_2.(j\omega)^2}{\frac{m}{R} + j\omega\frac{C_1 + C_3}{m} + (j\omega)^2 l_f \frac{m}{R}(C_1 + C_2) + (j\omega)^3 l_f \left(\frac{C_1C_2 + C_1C_3 + C_2C_3}{m}\right)} \qquad [1.29]$$

The capacitances C_1, C_2 and C_3 are expressed as:

$$C_1 = C_p + C_{ps}(1 - m) \qquad [1.30]$$

$$C_2 = m.C_{ps} \qquad [1.31]$$

$$C_3 = m^2 C_s + m(m-1)C_{ps} \qquad [1.32]$$

Two frequencies f_1 and f_2 are highlighted:

$$f_1 = \frac{1}{2\pi.\sqrt{l_f.m.(C_p + C_{ps})}} \qquad [1.33]$$

$$f_2 = \frac{1}{2\pi.\sqrt[3]{.\frac{l_f}{m}.(C_1.C_2 + C_1.C_3 + C_2.C_3)}} \qquad [1.34]$$

Calculating the transfer function was accomplished with typical values measured on a current transformer: C_p = 0.1 pF, C_s = 100 pF, C_{ps} = 1 pF, m = 100, R = 10 Ω, l_f = 50 nH.

For this example, the useful bandwidth at –3 dB stretches from 40 to 30 Hz. Beyond that, the parasitic effects begin to dominate: the series resonance lies essentially between the primary–secondary capacitance C_{ps} and the leakage inductance l_f (frequency f_1); then, for $f_1 < f < f_2$, we observe the dominant effect of the parasitic primary–secondary capacitance C_{ps} up until the parallel resonance frequency f_2. Beyond the frequency f_1, the current transformer is no longer usable.

Thus, the effects of primary-secondary capacitive coupling (capacitance C_{ps}) constitute a major limitation at high frequency, with the sensor acting as a high-pass filter. The result is the possibility of circulation of a common-mode current, prejudicial to measurements, as shown in Figure 1.39. Shielding between the primary and secondary of the device can limit these effects, with the parasitic capacitances C_{p1} and C_{p2} being in series. Finally, the cable connecting the sensor and the measuring tool can be influenced by high-frequency disruptive fields and by common-mode currents, which also translates into the deterioration of the measurement. To minimize these problems, it is preferable to insert the sensor into a conductor with low variability in order to limit the circulation of common-mode currents and to reduce the impedance from the shielding as much as possible.

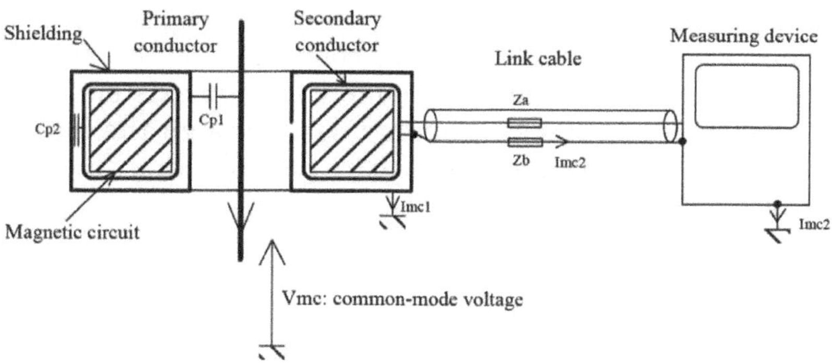

Figure 1.39. *Reduction of perturbations to the sensor through protective coating*

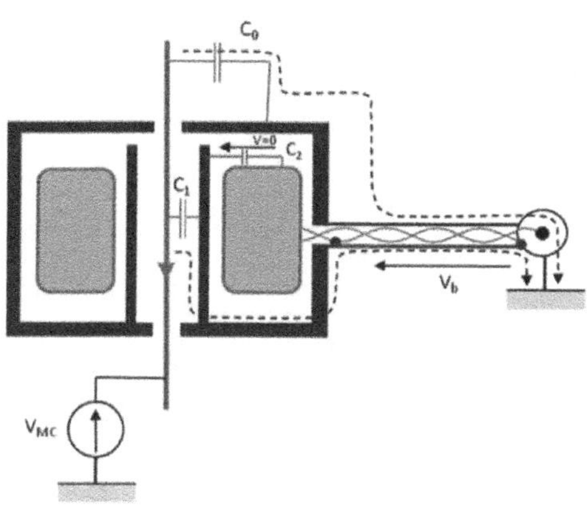

Figure 1.40. *Sectional view of a current transformer with double shielding*

In the case where we want a very strong immunity to common-mode currents, the alternative would be to use a shielded casing connected to a triaxial cable, as shown in the sectional view of the device in Figure 1.40. The parasitic common-mode currents resulting from either the conductor having a variable potential V_{MC} or any other source of parasitic electrical fields in the environment are transported via C_0 through the shielded case and the shielding of the triaxial cable, and likewise for the one due to part of the conductor being inside the casing via C_1. With the extremity of the secondary winding being connected to the shielded case, the current travelling through C_2 is non-existent. Thus, no common-mode current can circulate through the internal conductors of the cable that can perturb the secondary voltage of the transformer.

1.4.2.3. *Available passive current sensors and their specifications*

The sensors that currently exist on the market are limited to a frequency range of 4 or 5 decades, either toward the low frequencies (10 Hz–1 MHz) or toward the high frequencies (10 Hz–1 GHz). The curve in Figure 1.41 outlines the evolution of the high cutoff frequency with the current load, observed on existing products.

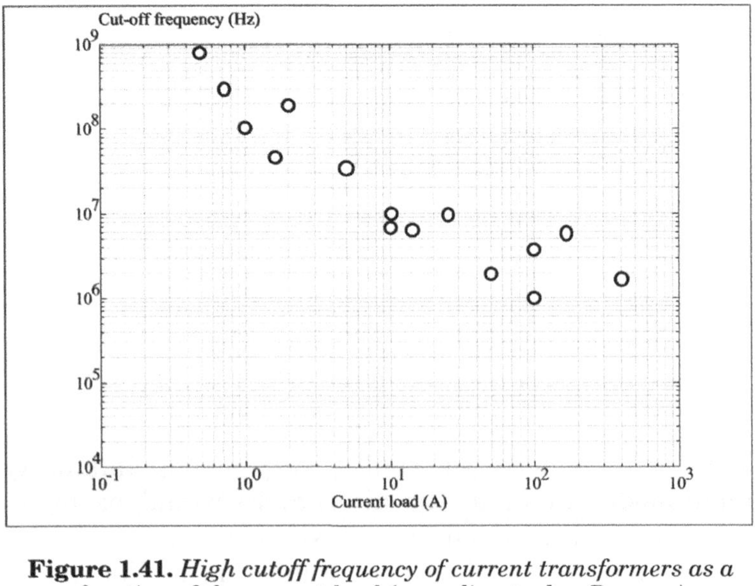

Figure 1.41. *High cutoff frequency of current transformers as a function of the current load (according to doc. Pearson)*

The specifications to take into account for choosing a sensor are therefore:

– the maximum current at different frequencies, generally at 50 Hz and the maximum usage frequency;

– the –3-dB bandwidth;

– the product $(I.T)_{max}$ that outlines the saturation during the application of a pulsed current;

– the maximum DC component before saturation;

– the value of the load resistance;

– the insertion impedance;

– the geometric and physical data (diameter of the aperture, bulk, weight, etc.).

1.4.3. *Antennae [KIN 82, BAL 97]*

Antennae are used for close-field (in other words for a distance $d < \frac{\lambda}{2\pi}$) or far-field measurements ($d > \frac{\lambda}{2\pi}$, λ is the wavelength of the perturbing signal). In this last case, we consider plane-waves. The wave impedance, defined by the ratio $\frac{E}{H}$, is constant and measures 377 Ω in far-field. Therefore, we only need to measure one of the two components in the electromagnetic field. The antennae for the far-field measurements are characterized by the following:

– The radiation pattern: this is the measurement of the amplitude E of the field emitted as a function of two angles defining the spherical coordinates of the space.

– The gain G, which is defined relative to an isotropic antenna emitting the same power:

G = maximum power density radiated in a specific direction/maximum power density uniformly radiated in every direction in space.

The gain of an antenna varies with the frequency; knowledge of this characteristic must be established for the frequency range.

– The antenna factor F_a: it is the relation between the field E and the voltage U measured at the ends of the antenna, which also varies with the frequency. It is established at a specifically predefined distance and is generally expressed in dB:

$$F_a = 20\log(\frac{E}{H})$$

The characteristics of antennae are adapted to the frequency range and the characteristics of the field that we want to measure. These characteristics must be as constant as possible in the frequency band being used. The references [INT 03a, INT 03b] define the characteristics of the antennae used in the measuring of fields. We will cite several currently used models.

– Band A (10–150 kHz), field H: observations show that it is the magnetic field that is responsible for the perturbations. The antenna is a loop in an electrically shielded frame and must fit within a square of 0.6 m.

– Band B (150 kHz–30 MHz), field H: we use the same device as in band A; field E: we use a vertical 1 m whip antenna for measuring a distance $d < 10$ m.

– Band C (30–300 MHz), field E: we use a balanced doublet antenna of length $\frac{\lambda}{2}$ for $F > 80$ MHz (λ corresponding to $F = 80$ MHz that being 3.75 m). Its symmetry is such that the measured level varies by at least

20 dB when its orientation passes from the parallel polarization direction to the transverse polarization direction. It is possible to use short doublet antennae (however their length is limited to $\frac{\lambda}{10}$), averaging the knowledge of their antenna factor. They are less sensible however. A frequently used model is the biconical antenna (Figure 1.42) that aims to present a large frequency range.

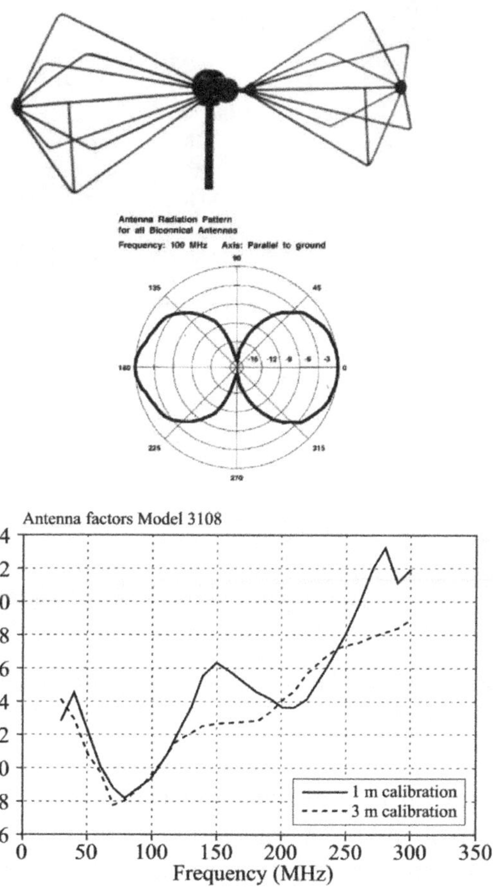

Figure 1.42. *Biconical antenna and its antenna factor curve (according to doc. EMCO)*

– Band D (300–1,000 MHz): the antenna must be polarized in a plane. We use more complex antennae than the doublet, for example the log-periodic antenna, made up of coupled components, whose tuning frequencies are in a geometric progression, which gives it a large bandwidth. An example is given in Figure 1.43.

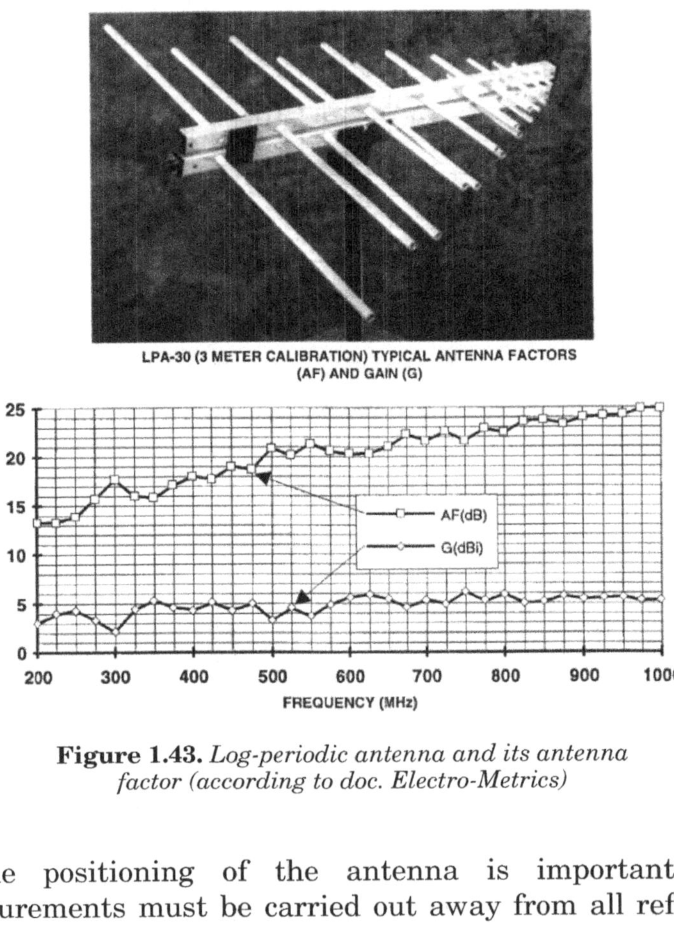

Figure 1.43. *Log-periodic antenna and its antenna factor (according to doc. Electro-Metrics)*

The positioning of the antenna is important. The measurements must be carried out away from all reflective surfaces. The height relative to the ground and the distance relative to the perturbation source must be known. The preferable distances are 3, 10, 30 m, etc. We can consult [INT 03a] for precise recommendations regarding radiated measuring techniques.

1.4.3.1. *Magnetic near-field probes*

When it is necessary to locally characterize an electromagnetic field close to its emission source, we are in the context of near-field measurements. For magnetic near-field measurements, there are two major types: Hall probes and loop probes.

1.4.3.2. *Hall probes*

The probes presented here use the Hall effect: if conductive tape is traversed by an electric current I (along the x-axis) and if a magnetic field $\vec{B}$ is perpendicular to the plane of the tape (along the y-axis), an electric field $\vec{E}$ appears and therefore so does a difference in potential between the edges of the tape; this phenomenon is called the Hall voltage. This voltage is oriented according to the direction perpendicular to these two axes (along the z-axis), with its orientation visible in Figure 1.44.

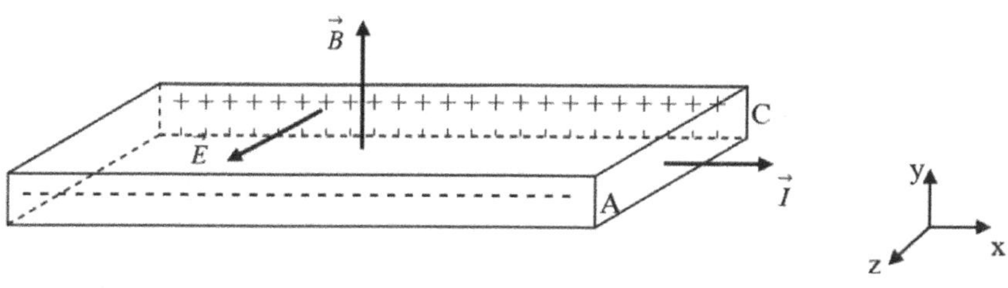

Figure 1.44. *Hall probe*

Hall probes help measure not only static magnetic fields, but also alternative magnetic fields up to approximately 100 kHz. Nevertheless, the Hall voltage varies with time and temperature depending on the fabrication of the probe, which requires frequent calibration. These are of a low dimension and high dynamics. They help in measuring the magnetic field within a range of 100 µT–10 T, their usage being quite restricted in EMC measurements.

1.4.3.3. *Loop antennae*

The magnetic field can be measured using circular or rectangular loop probes [WHI 64]. The principle of this probe rests on the immersion of a coil (one or more loops) of various forms (round, square, trapezoid, etc.) into a magnetic field. The circular probe is one of these probes. By considering a circular loop whose diameter is small compared to the wavelength, subjected to an electromagnetic plane wave where the B component is perpendicular to the plane of the loop, the E component will be in the plane of this loop. This loop is loaded by an impedance Z_c as we can see in Figure 1.45. Z_c is the impedance of the load of the measuring sensor, generally 50 Ω; we note that this connection dictates that one of the ends of the antenna must be earthed via the measuring device. This is not without consequences as we will see further on.

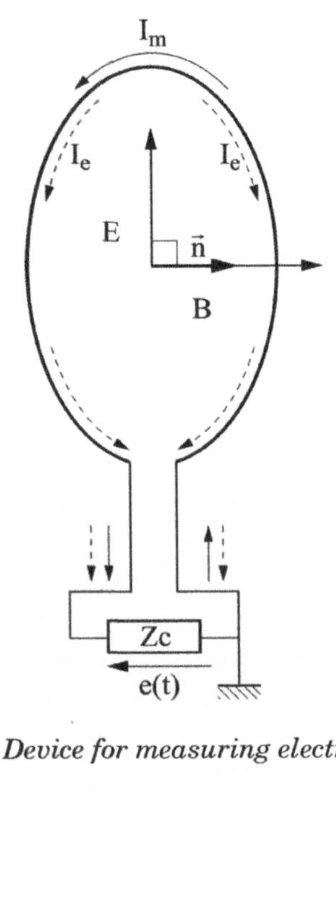

Figure 1.45. *Device for measuring electromagnetic field*

Each component in the field creates an EMF in the loop, where the B component induces a voltage spread around the turn; the E component induces a voltage spread in the direction of E in the plane of the loop (in accordance with Figure 1.45). The result is the circulation of two types of current I_m and I_e due, respectively, to B and E, their combined effects making up the voltage e(t) at the ends of Z_c. Indeed, the connection to earth allows the circulation of a displacement current due to the vertical component E of the field in each half-branch of the loop: the loop behaves like an electric monopole, the effect of which is more sensible the higher the frequency.

A simplified electrical diagram is given in Figure 1.46 [KAN 93]. There are two electromotive forces: e_m generated by the magnetic field B and e_e generated by the magnetic field E. Each electromotive force is in series with a radiation impedance Z_{re} or Z_{rm} as well as the inductive and capacitive resistive elements in the loop: R_a, L_e, C_e, L_{a0}. The current I_a circulating in the impedance Z_c is the sum of the two currents I_e and I_m that are circulating in the electrical and the magnetic branches of the circuit, respectively.

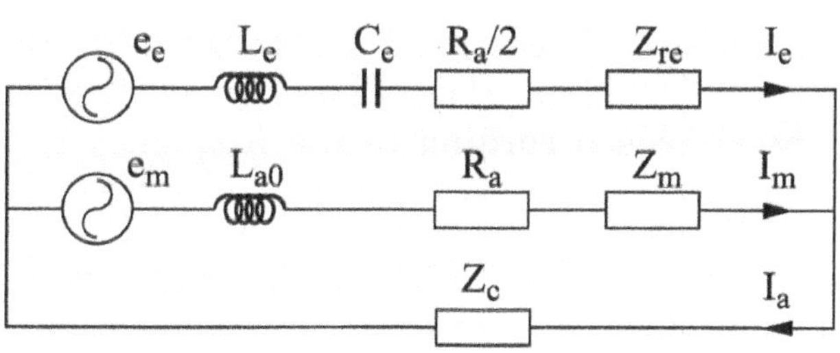

Figure 1.46. *Electrical diagram of a loop probe*

The electromotive force e_m of the magnetic side is expressed here, with A the surface and D the diameter of the turn:

$$e_m = -j2\pi fBA \qquad [1.35]$$

$$e_e = \pi \frac{D}{2} E \qquad [1.36]$$

L_{a0} is the low-frequency inductance of the loop and d is the diameter of the conductor making up the turn, which are expressed as:

$$L_{a0} = \mu_0 \frac{D}{2}\left(\ln\left(\frac{8D}{d}\right) - \frac{7}{4}\right) \qquad [1.37]$$

The resistance R_a is the ohmic resistance of the loop, which is defined in relation to the resistivity ρ and the section s such that:

$$R_a = \frac{\rho \pi D}{s} \qquad [1.38]$$

In the magnetic branch of the diagram in Figure 1.45, this resistance intervenes in its entirety, but only half in the electric branch. We note that the radiative impedance value Z_{rm} is negligible for a single-turn loop antenna [BAL 97]: for example, for a brass antenna made up of a single turn of wire with a 1 mm section and a 2 cm diameter, at 100 kHz, the value of R_a is 4.3×10^{-4} Ω while the value of Z_m is 3.8×10^{-17} Ω. In practice, we can see that for D > λ/10, the electric side becomes negligible. The diameter of the probe can therefore be shaped according to the frequency range being used.

In the event that we are looking to measure only the B component of the electromagnetic field over a large frequency range (up to several tens of megahertz), it becomes necessary to shield the probe. This is achieved by winding the measuring turn(s) with a shielding that is open at the extremity opposite the measuring charge (necessary in relation to the B component), as shown in Figure 1.47. The voltage e_e is induced exclusively on the shielding and affects neither the frequencies on the turn nor the measuring impedance.

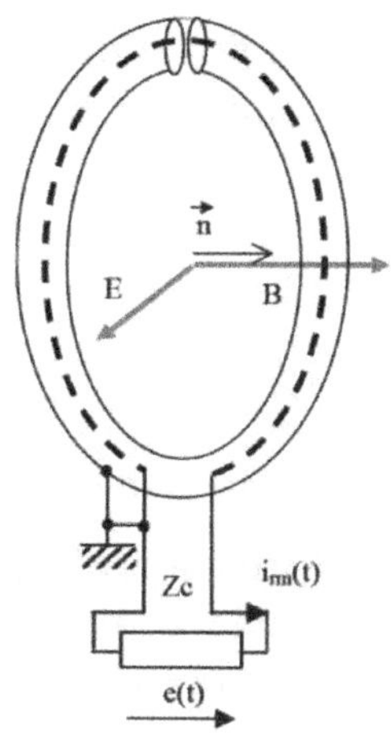

Figure 1.47. *Shielding of the measuring loop*

Thus, the shielding of the probe helps to minimize the impact of the electric field on the measurement of the magnetic field. We will therefore take note of the field probe model shown in Figure 1.48, relating to the quantities defined above.

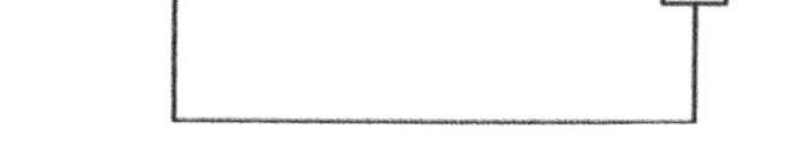

Figure 1.48. *Simplified model of a field probe*

The frequency transfer function of the probe is given as:

$$T_B(\omega) = \frac{e(\omega)}{B(\omega)} = \frac{Z_c j\omega\pi\frac{D^2}{4}}{Z_c + R_a + j\omega L_{a0}} \qquad [1.39]$$

It is a high-pass function for which the chopping pulsation ω_c is given by (supposing that $Z_c = R_0 = 50\ \Omega$ and that $R_0 >> R_a$):

$$\omega_c = \frac{2R_0}{\mu_0 D\left(\log(\frac{8D}{d}) - \frac{7}{4}\right)} \quad [1.40]$$

1.4.3.3.1. Compromises on the size of the probe

If we substitute the expression of diameter D of the probe into [1.35] instead of its surface A and it is made of *n* turns to improve its sensitivity, we obtain:

$$e = -j2\pi f \cdot n \cdot B\pi \frac{D^2}{4} \quad [1.41$$

The sensitivity of the probe is given by:

$$\left|\frac{e}{B}\right| = 2\pi f\pi \cdot n\frac{D^2}{4} = \pi^2 f \cdot n\frac{D^2}{2} \quad [1.42$$

It is simpler for studying purposes to consider the spectral sensitivity, with [1.42] becoming:

$$\left|\frac{e}{Bf}\right| = \pi^2 n\frac{D^2}{2} \quad [1.43]$$

It is therefore easy to see that the bigger the diameter and number of turns, the better the sensitivity of the probe. Thus, to reduce the dimension of the antenna and obtain a more precise measurement, it is preferable to decrease its diameter; nonetheless, this results in a lower sensitivity and, if we increase the number of turns, degrades its bandwidth through the increase in parasitic capacitive effects.

Numerical Example

We wish to measure a minimum magnetic induction B of 10 nT (or 8×10^{-3} A/m) at 10 kHz. With the minimum voltage sensitivity of the measuring device being 1 µV, we deduce that for one turn, the diameter is given by:

$$D = \frac{1}{\pi}\sqrt{\frac{2e}{nBf}} = 45\,\text{mm} \qquad [1.41]$$

To illustrate this, the following table gives the characteristics of a commercial probing kit (*doc.* Ets-Lindgren). They are made up of a semi-rigid coaxial loop soldered to itself to make a turn; an opening is kept in the shielding to ensure the penetration of the field B.

Diameter	**Impedance**		
	R	**L**	**Module**
6 cm	237.8 mΩ	203.24 nH	12.77 Ω
3 cm	135.7 mΩ	103.1 nH	6.48 Ω
1 cm	99.46 mΩ	46.79 nH	2.94 Ω

Table 1.6. *Values of parameters for the loop probes*

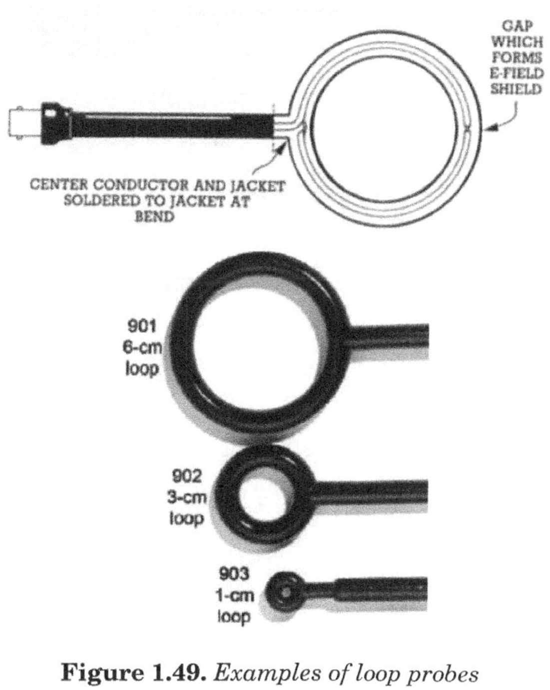

Figure 1.49. *Examples of loop probes*

Example of application

A near-field probe can be used to create a local mapping of the field B, with this analysis helping to determine the areas where the emitted field is susceptible to perturb neighboring operations in an appliance or even to characterize the radiating circuit as an equivalent source. This is illustrated in Figure 1.51 where we have characterized through mappings of two differing frequencies the radiating H-field vertical component of the chopper represented in Figure 1.50. The measurement was made at switching frequency (20 kHz); we can clearly see the heavy radiation of the switching cell, which helps to determine its total parasitic inductance (L_{mesh} ~ 450 nH) by integrating the measured field over the measuring surface:

$$L_{maille} = \frac{\mu_0}{I} \iint_{surface} H_z(x,y)dxdy \qquad [1.42]$$

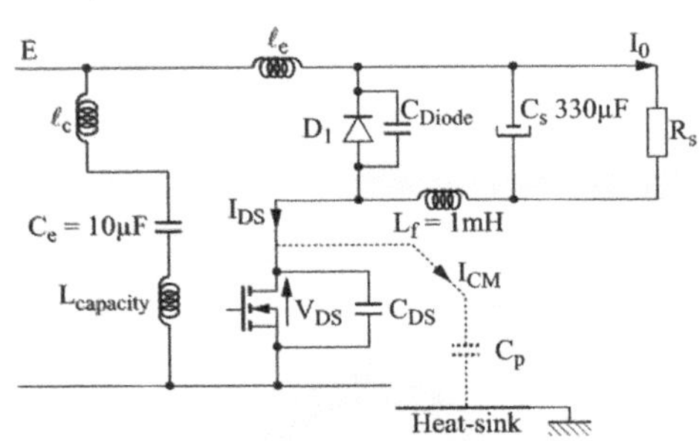

a) Electrical diagram of a step-down chopper

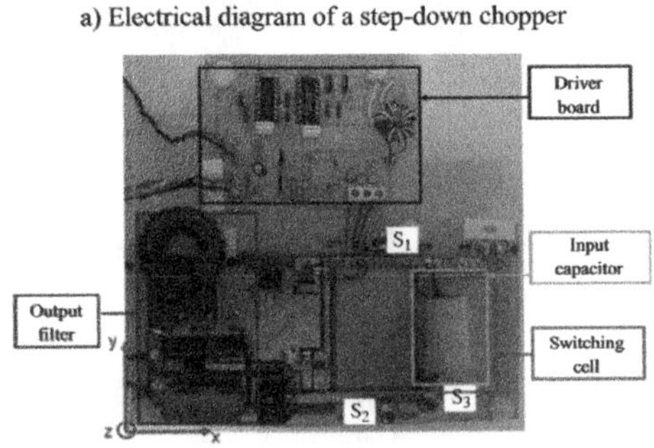

b) Experimental device

Figure 1.50. *Diagram and view of a chopper*

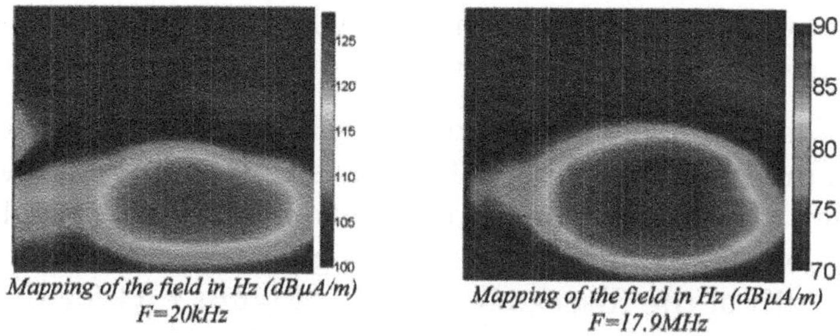

Figure 1.51. *Mappings of magnetic fields near the chopper. For a color version of this figure, see www.iste.co.uk.costa / electronics.zip*

We can also observe a significant radiation at 17.9 MHz but with less amplitude in the same zone due to HF resonances during switchings. The exploitation of this kind of measurement can improve the design of electronic circuits through a better understanding of the radiated effects in near-field.

1.4.4. *Spectrum analyzer*

This is the device that is preferably used for measurements in EMC. It is used to analyze the properties of perturbation signals or to carry out measurements while following standardized procedures. Its simplified structure is represented in Figure 1.52.

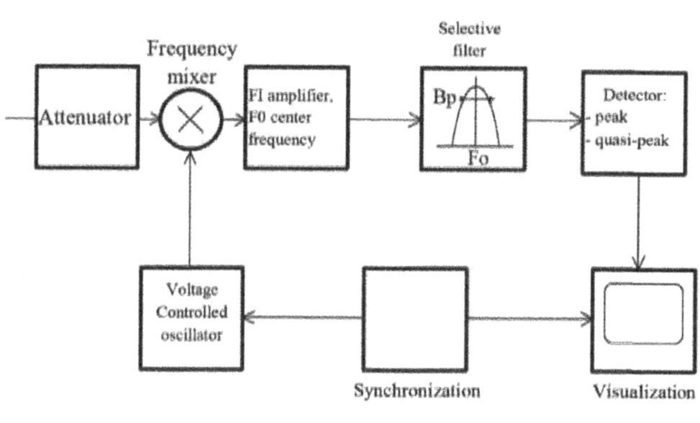

Figure 1.52. *Structure of a heterodyne spectrum analyzer*

The main points to consider when carrying out EMC measurements are the following:

1) *The frequency resolution*: this is the bandwidth of the analyzing filter. It ensures that measurements are made in a narrow band (only one signal harmonic in the resolution band) or large band (several signal harmonics in the bandwidth of the filter), as shown in Figure 1.54 for the case of a repeating perturbation signal, thus having a peak spectrum. The narrower the bandwidth of the filter is, the longer the measurement time will be, and the signal-to-noise ratio (S/N) will be greater. Thus, the sweeping speed v (in ms/cycle) is determined through the expression:

$v \leq B_p^2$ B_p is the bandwidth at –6 dB of the analytical filter.

The choice of the resolution bandwidth (RBW) also affects the S/N of the measurement, which increases by 10 dB when the RBW decreases by a factor of 10, as seen in Figure 1.53.

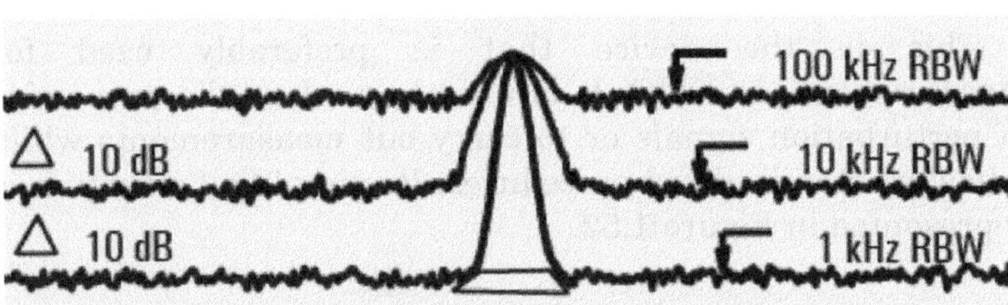

Figure 1.53. *Evolution of the signal / noise resolution as a function of the resolution bandwidth of the analyzing filter*

Impulsive or very low-recurrence signals are always measured in large band. Standards define the resolution according to the measured frequency band, as shown in Table 1.7.

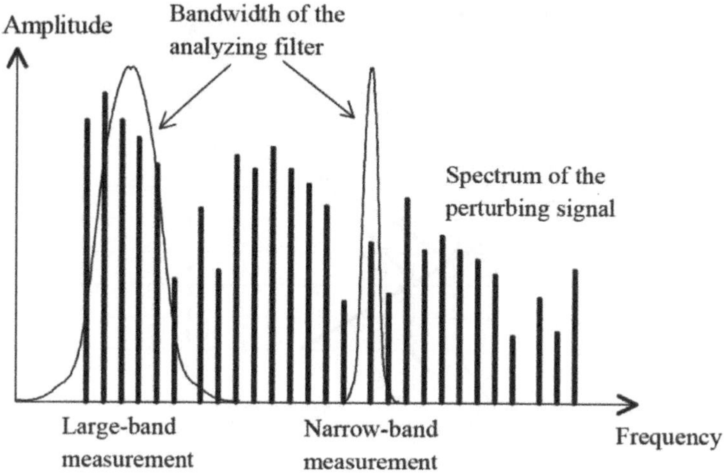

Figure 1.54. *Signal measured in narrow band and large band*

Frequency range	A	B	C and D
Bandwidth (–6 dB)	200 Hz	9 kHz	120 kHz

Table 1.7. *Standardized resolution bandwidths for EMC measurements*

2) *Input attenuator and mixer*: the mixer is a nonlinear device that introduces two kinds of errors on the display:

i) Its gain decreases when the amplitude of the signal increases (compression). This error is observable by adjusting the input attenuator upwards until we can read a level that is unrelated.

ii) It introduces distortion; the makers distinguished second- and third-order harmonic levels on the output signal depending on the input level.

Figure 1.55 shows the evolution of the S/N as a function of the input power at the level of the mixer with the previously mentioned limitations. It is therefore important to adjust the input attenuator so that the measurement is carried out with the best possible S/N.

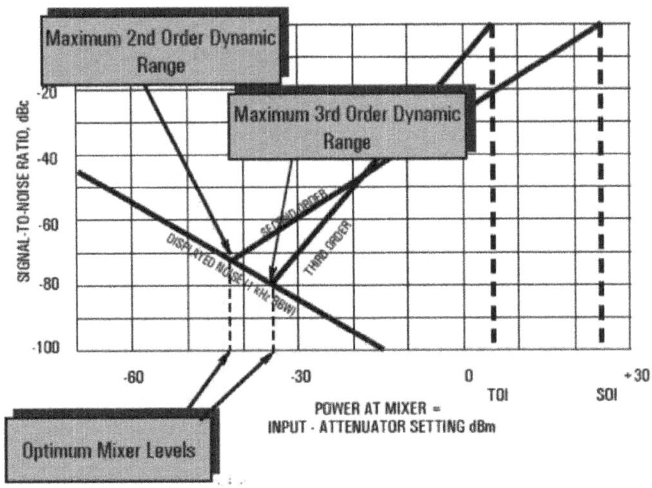

Figure 1.55. *Evolution of the signal / noise ratio as a function of the input power (doc. Agilent)*

Nevertheless, the attenuator introduces noise into the measurement, and the S/N degrades by 10 dB when the attenuation increases by the same amount.

3) *Detection methods*: the main detection methods are the peak, quasi-peak (QP) and average detection modes. Standards recommend QP detection as it is well adapted to the quantification of the acoustic inconvenience that results from radiofrequency perturbations.

1.4.4.1. *The QP detector*

The QP detector was introduced in the 1930s to measure the effects of interferences in the reception of signals broadcasted in amplitude modulation (AM). With a large part of these perturbations being of an impulsive nature, and with the effects of the interferences increasing with the repetition frequency, the QP detector connected to a low-frequency voltmeter ensures a measurement that is representative of the trouble of listening to the broadcast. The time constants of the QP detector were defined by

CISPR, initially for a frequency range of 150–160 kHz, which corresponds to the AM radio band, before being extended to 30 MHz[4].

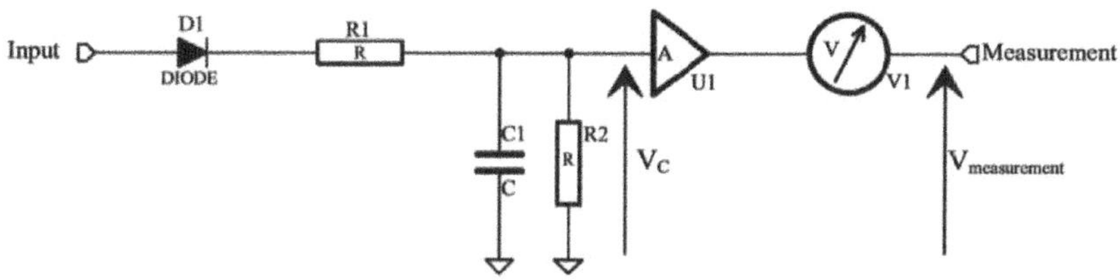

Figure 1.56. *Structure of the quasi-peak detector*

The structure of the QP detector is represented in Figure 1.56. The time constant is defined by $T_C = R_1//R_2.C_1$ and the discharge time by $T_D = R_2.C_1$. The voltmeter at the output acts as second-order critical damping, the time constant of which is given in Table 1.8.

Characteristics	9–150 kHz (Band A)	150 kHz–30 MHz (Band B)	30 MHz–1 GHz (Band C)
Bandwidth of the analytical filter (–6dB)	200 Hz	9 kHz	120 kHz
Charge time (Tc) (ms)	45	1	1
Discharge time (Td) (ms)	500	160	550
Time constant of the critical damping voltmeter (ms)	160	160	100

Table 1.8. *Characteristics of the quasi-peak detector as a function of the frequency, standards CISPR 16-1-1 and ANSI C63.2*

4 "The Quasi-Peak Detector", by Edwin L. Bronaugh, ANSI ASC C63 Historian.

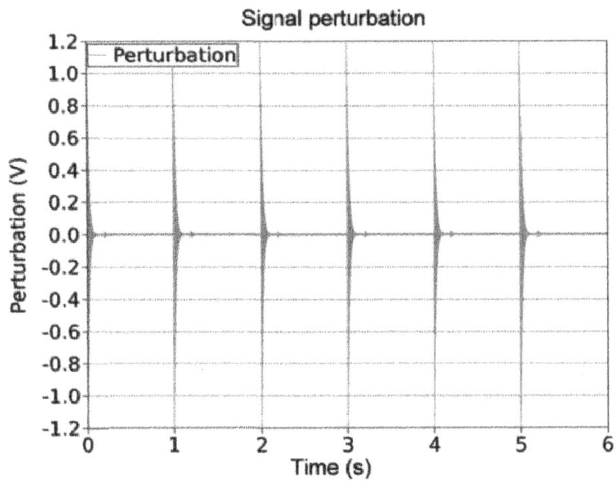

Figure 1.57. *Parasitic signal, repetition frequency 1 s, analytical filter frequency 50 kHz*

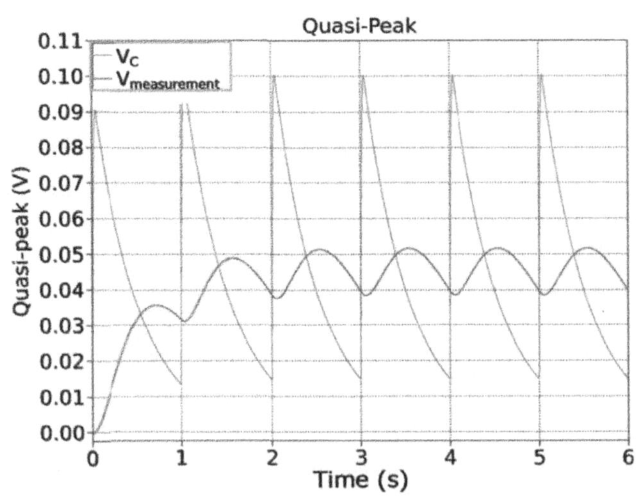

Figure 1.58. *Example of band A quasi-peak detection*

The output level of the QP detector varies with the frequency of the perturbation. This detection method results in much longer detection times than peak detection. The average detection value provides a level proportional to the signal. Figure 1.59 shows the levels provided by the three detectors as a function of the perturbing signal frequency, with standardized tuning specifications.

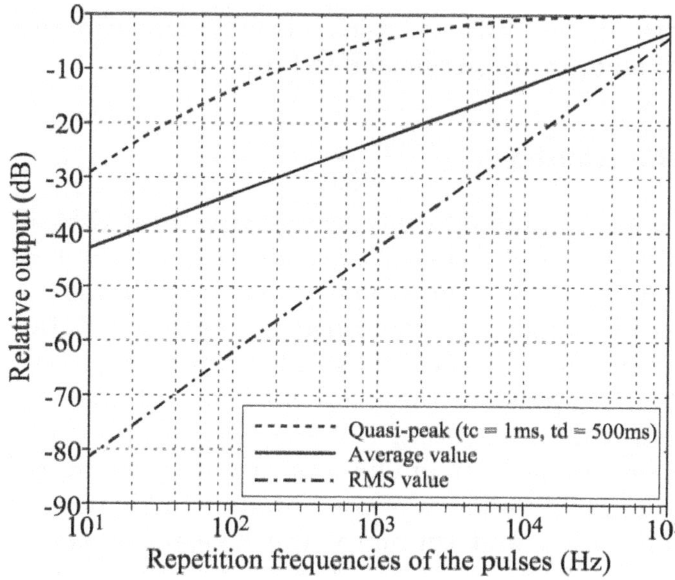

Figure 1.59. *Output levels for different detectors as a function of the disruption frequency*

The observations show that the measurement of a signal can vary greatly depending on whether it is in a narrow band or large band and on the detection method. A power electronic device can operate with several different switching frequencies (rectifier at 100 Hz, inverter at 20 kHz).

If the goal is to distinguish the sources of noise, spectral analysis must then be performed in narrow band with a resolution filter inferior to the lowest switching frequency. The detection method must be the peak type, as it is the quickest and best adapted.

On the other hand, if the goal is to carry out standardized measurements, the choice of the analysis method is imposed by the recommendations of CISPR. The measured levels can therefore be very different to what is observed in the previous case, and not reflect the true perturbing power of the device, as this is particularly overestimated if the measurements are performed in large band.

1.5. The standards

Two main standards categories exist: those that define the tolerated levels of radiated or conducted emissions and those that define the electromagnetic susceptibility of an appliance, namely its tolerance to electromagnetic strikes. The main objective of these standards is to outline as precisely as possible the conducted or radiated measuring environment (measuring in free space, in an anechoic chamber, support for appliances), the measuring conditions (cable lengths, distance of the antennae, height, angle, etc.), the calibration and tuning of the measuring devices being used, all with the goal of carrying out easily reproduced and reliable measurements. Since January 1, 1996, all electrical equipment have had to conform to one or more EMC standards within its category. Traditionally, the standards were established for the protection of broadcasting [COO 79]; however, this field has now expanded as the susceptibility standards can testify [INT 03b]: resistance of a particular equipment category to electrostatic discharges, to induced currents, to electromagnetic fields, to nuclear EMP (electromagnetic pulse). We will end here by mentioning emission standards concerning equipment using static converters. Table 1.9 summarizes the main standards. We must take into account that the last line corresponds to grid harmonics, which fall into the category of low-frequency perturbations conducted in power networks.

European reference	Field of application	International standard
EN 55011	Measuring methods and limits for electromagnetic perturbations in industrial and scientific appliances and medical electric radio frequency equipment.	CISPR 11
EN 55014	Measuring methods and limits for electromagnetic perturbations in electric household appliances and electric portable equipment, related to radio frequencies.	CISPR 14
EN 55015	Measuring methods and limits for electromagnetic perturbations in fluorescent lamps and electric lighting, related to radio frequencies.	CISPR 15
EN 55022	Measuring methods and limits for radio frequency perturbations in information processing machines.	CISPR 22
EN 61000-2-3	Harmonic perturbations produced in power-supply networks by electric household appliances.	IEC 61000-2-3

Table 1.9. *Major standards involving static converters*

1.6. Bibliography

[BAL 97] BALANIS C.A., *Antenna Theory – Analysis and Design*, 2nd ed., John Wiley and Sons, Hoboken, 1997.

[COO 79] COOK J.H., "Quasi-peak-to-RMS voltage conversion", *IEEE Transactions on Electromagnetic Compatibility*, vol. EMC-21, no. 1, pp. 2–9, February 1979.

[GUE 02] GUENA A., COSTA F., LABARRE C., GAUTIER C., "Unified technique in power electronics: the thin wire method", *4th International Conference on Computation in Electromagnetics (CEM 2002)*, Bournemouth, UK, 2002.

[INT 03a] INTERNATIONAL SPECIAL COMMITTEE ON RADIO INTERFERENCE, "CISPR 16-1, Specification for radio disturbance and immunity measuring apparatus and method", Commission, 2003.

[INT 03b] INTERNATIONAL SPECIAL COMMITTEE ON RADIO INTERFERENCE, "CISPR 16-2, Specification for radio disturbance and immunity measuring apparatus and method", Commission, 2003.

[KAN 93] KANDA M., "Standard probes for electromagnetic field measurements," *IEEE Transactions on Antennas and Propagation*, vol. 41, no. 10, pp. 1349–1364, 1993.

[KIN 82] KING RONOLD W.P., "The cylindrical dipole as a sensor or probe", *IEEE Transactions on Electromagnetic Compatibility*, vol. Emc-24, no. 3, pp. 364–367, August 1982.

[PAU 92] PAUL C.R., *Introduction to Electromagnetic Compatibility*, Wiley Interscience, 1992.

[UMA 87] UMAN M.A., *The Lightning Discharge*, vol. 39 of the International Geophysics Series, Academic Press, Orlando, FL, 1987.

[WHI 64] WHITESIDE H., KING R., "The loop antenna as a probe," *IEEE Transactions on Antennas and Propagation*, vol. 12, no. 3, pp. 291–297, May 1964.

Chapter 2

Fundamental Principles

2.1. Sources of noise: the switching cell and its control

The static conversion of electrical energy (switch-mode power supply, inverter, rectifier, etc.) is based on the principle of the switching cell: it is the connection of two switches, which enables the management of energy between an input source of voltage and an output source of current (Figure 2.1). The main switch is controlled by a periodic modulation function *fm*(t), with a binary value and a variable cyclic ratio ($\alpha = t_{on} / T_d$), which regulates the transfer of power according to the value of this cyclic ratio α. The notion of source must be understood in the sense that it is capable of imposing a near-constant quantity (of voltage or current) at the time-scale of the switching period. This characteristic is generally due to the presence of reactive components (input capacitor C_e or output smoothing inductance).

Thus, it is noteworthy to see that the external parameters of the switching cell (E and I_o) are constants, whereas the internal parameters (i_e and v_k) are variables, regulated by the function *fm*(t).

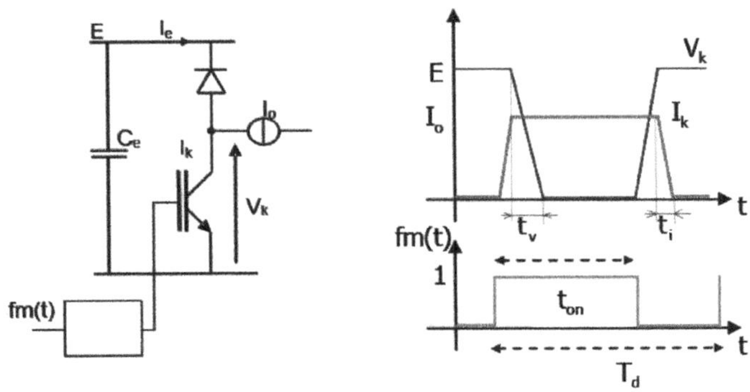

Figure 2.1. *Switching cell and its corresponding wave shapes*

It is possible to express the variable voltages and currents of the switching cell as a function of those that are constant:

$$V_k(t) = fm(t) \cdot E \quad [2.1]$$

$$I_e(t) = fm(t) \cdot I_o \quad [2.2]$$

From these relations, we deduce the law of converted power control:

$$P_s = \langle v_k \rangle \cdot I_o = \alpha \cdot EI_o \quad [2.3]$$

Thus, by this principle, all static conversion functions can be undertaken. We sometimes add a transformer to the structure when a galvanic insulation is required.

2.1.1. *Origin of conducted and radiated perturbations in static converters*

The electrical quantities are very much variable in the switching cell. Indeed, in order to reduce losses during commutation (simultaneous presence of the voltage and the current in the switches), it is essential that the commutations be very quick. Currently, the size of switching gradients is on the scale of 100–1,000 A/µs for dI/dt and of 5–50 kV/µs for dV/dt.

To illustrate, Figure 2.2 outlines these phenomena in a chopper connected to a line impedance stabilization network (LISN): in the mesh surrounding the hatched area, the current I_e undergoes very quick high-frequency variations. The resulting loop connects to a magnetic radiative dipole: the input decoupling capacitor C_e, limited in its operation due to its imperfections (resistance and inductance in series l_p), is generally not sufficient to prevent the propagation of an impulsive parasitic current I_p onto the network.

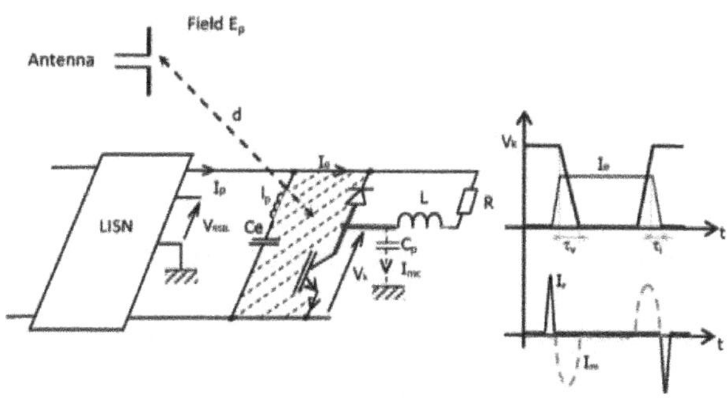

Figure 2.2. *Origin and coupling mode of electromagnetic perturbations of a static converter*

Moreover, the conductors shown in grey bold lines in Figure 2.2 endure the strong variations of the voltage V_k. They constitute an electric radiative dipole and can transmit impulsive currents I_{mc} to the earth via the parasitic capacitance denoted by C_p between the device and the earth.

2.2. Modeling

2.2.1. *Simple model of the switching cell*

We can now provide a model of the switching cell, representing the effects of perturbation [COS 93]. The input current of the cell is presented as a current generator creating the parasitic differential-mode current by means of coupling through a common impedance (input capacitor). The voltage of the switch is presented as a voltage generator generating the common-mode current via capacitive couplings. The switching cell can then be represented by one of the two models in Figure 2.3, where the sources of current I_e and voltage $V_{mc} = V_k$ appear.

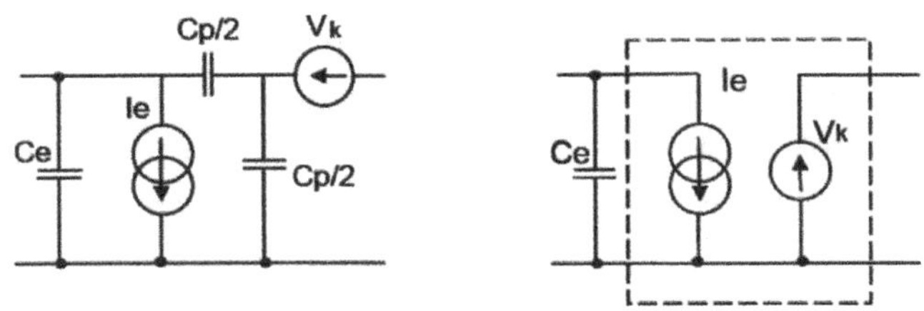

Figure 2.3. *Models of the switching cell for the depiction of conducted perturbations*

These models can be connected with that of the electric environment of the converter (network, charge, control devices and connections to the earth) to determine and calculate the common- and differential-mode currents at the input while conforming to the standards as required.

The first model shows why it is important to see that the parasitic currents due to the voltage V_k can flow along the two power supply lines, but it also includes the common-mode coupling capacitances. The source of voltage V_k is in a floating reference. The second model is linked to the common reference voltage, which simplifies the analysis, and is more representative of the switching cell. These equivalent generators are, therefore, considered as equivalent sources of electromagnetic (EM) perturbations in the frequency domain. We complete the description of the model with the impedances linking the different potentials of the cells to the potential reference (earth, frame, etc.). These intangible impedances represent the effects of parasitic couplings between the converter and its electrical environment. All of the imperfections of the active and passive components can be included in this model (Figure 2.4). This representation, therefore, suggests that the effects are local, which is an acceptable hypothesis as long as the geometric dimensions of the devices remain small relative to the wavelengths of the quickest signals, which is generally the case.

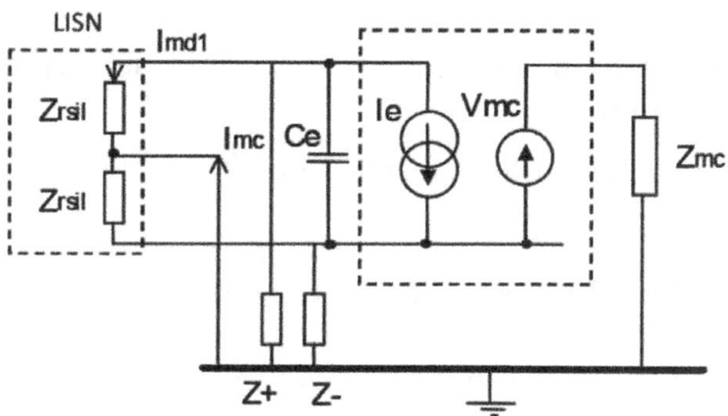

Figure 2.4. *Electromagnetic compatibility (EMC) models of the switching cell through equivalent sources and local couplings*

By completing this model with the model incorporating the LISN, it is easier to carry out the calculation of the current I_p applied to the measuring impedance of the LISN, and representative of the contribution of each source to the disruptive signal. We can, therefore, express the current I_p that circulates through each impedance of the LISN as:

$$I_p = I_e \cdot C_1(f) + V_k \cdot C_2(f) \qquad [2.4]$$

The functions $C_1(f)$ and $C_2(f)$ are representative of the source couplings I_e and V_k in differential mode and common mode in the frequency domain. In fact, they are analogous to transfer functions that help to connect the internal perturbation sources to the quantities measured by the LISN. These functions can be calculated analytically by localizing all of the impedances of the system. They can also be obtained experimentally by directly measuring with the impedance analyzer. We can, therefore, see that the sources present in the converter contribute to the conducted perturbations. Nonetheless, certain coupling functions are dominant relative to each propagation mode; in particular, the function C_1 is dominant in differential mode as is the function C_2 in common mode. Two points, therefore, arise as determiners:

– the excitation sources I_e, V_k, characterized in the frequency domain;

– their coupling functions $C_1(f)$ and $C_2(f)$.

A very simplified example of execution is illustrated in Figure 2.4. For this scenario, we have presented the coupling functions $C_1(f)$ and $C_2(f)$ (Figure 2.5). We can note the minimum of C_1 due to the self-resonance of the capacitor C_e.

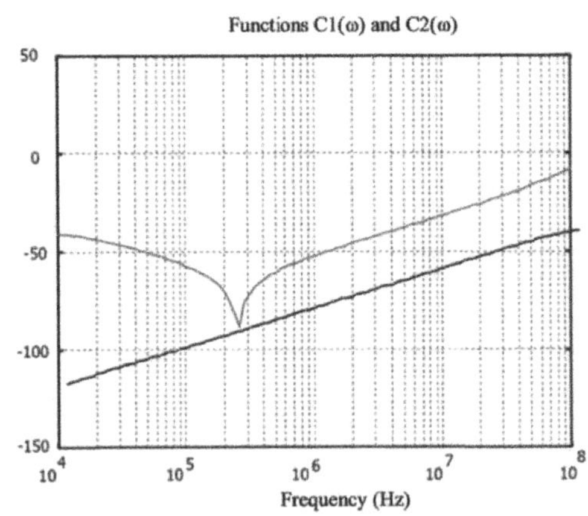

Figure 2.5. *Evolution of the frequency of the coupling functions $C_1(f)$ and $C_2(f)$*

We can deduce the spectrum of conducted perturbations in the LISN through the application of expression [2.4]. Figure 2.6 shows a comparison between the calculated and measured spectra by using this modeling principle. The support is a chopper switching at 15 kHz, supplied at 50 V and providing an output current of 2 A. We will reintroduce it later on.

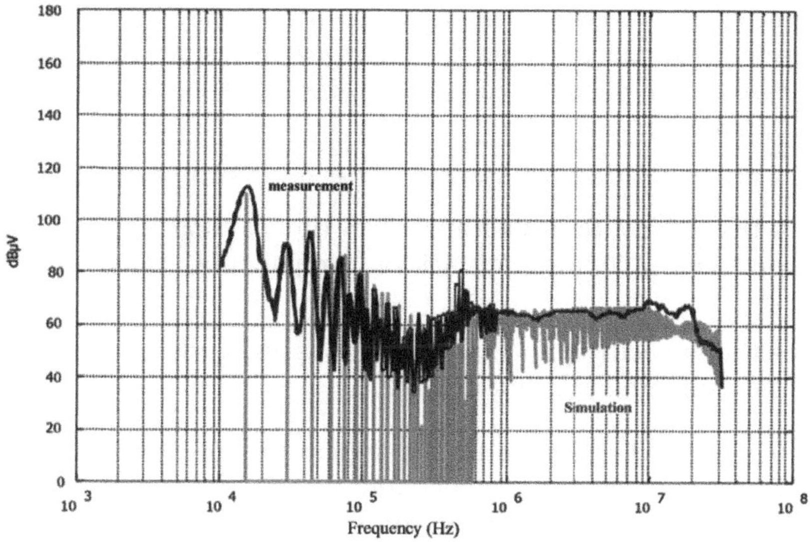

Figure 2.6. *Comparison between a simulation based on the frequency model and measurements for a scenario including a chopper*

This approach can also be applied to the radiated mode: the coupling functions are much more complex as they integrate the (generally three dimensional (3D)) geometric properties of the source circuits and the coordinates of the observation point of the radiation. Nevertheless, we can express the radiated fields as alternative equivalent forms, which means that the coupling functions are dependent on the nature of the observed field:

$$E_p = I_e \cdot R_1^E(\textit{geometric quantities, circuit}) + V_k \cdot R_2^E(\textit{geometric quantities, circuit}) \quad [2.5]$$

$$H_p = I_e \cdot R_1^H(\textit{geometric quantities, circuit}) + V_k \cdot R_2^H(\textit{geometric quantities, circuit}) \quad [2.6]$$

In every scenario, it is the electric parameters of the switching cell that must be considered as the noise sources of the converter.

2.2.2. *More complex model of the switching cell*

The model of the sources defined earlier do not take into account the presence of very high-frequency rays in the spectrum of parasite signals. Their origin resides in multiple parasitic components in the conversion structure (Figure 2.7), namely:

– the parasitic capacitances of the open-state switches;

– the parasitic inductances of the cabling or inherent components (bonding of active components, capacitor C_e, etc.);

– the capacitive and inductive couplings with the ground of the switching cell and its load.

Their effect is shown by the appearance of high frequency (HF) resonances in the spectrum of parasitic signals, which become greater the quicker the switch is controlled, as the switching losses are barely dampened in idle states. These systems can attain several tens of megahertz, and even more so for low-voltage high-current converters.

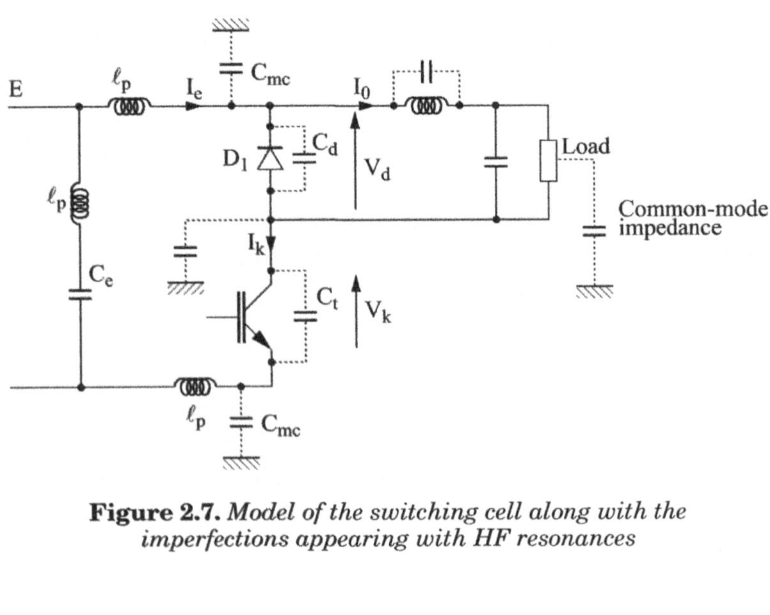

Figure 2.7. *Model of the switching cell along with the imperfections appearing with HF resonances*

To illustrate these effects, Figure 2.8 shows the current measured in the chopper being studied in the previous section. We will see parasitic HF resonances during commutations involving the major parasitic elements: total parasitic impedance of the switching cell and the capacitances of the semiconductors when they are in open state.

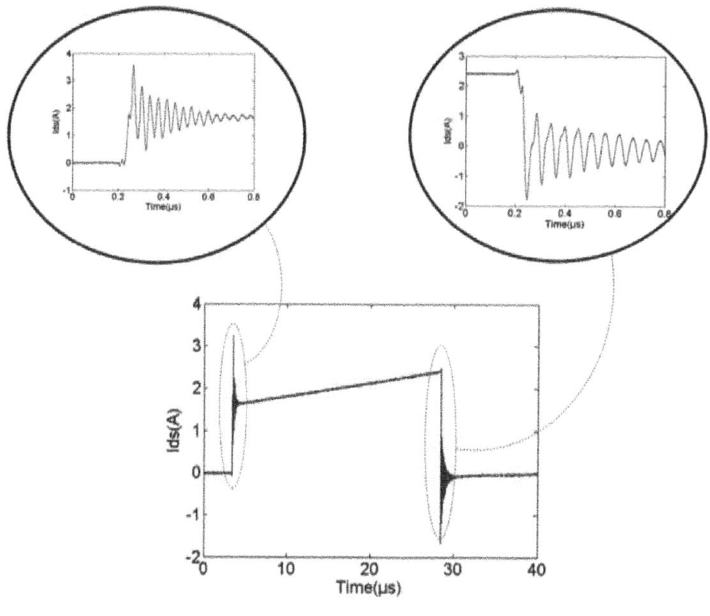

Figure 2.8. *HF oscillation states during commutation of the power transistor in a chopper*

If we consider an instantaneous commutation of the switch, and by noting that the damping m is inevitably very weak, the expressions of the parameters are defined as such (the expressions are simplified):

– ℓ_t : sum of all the parasitic inductances of the switching cell;

– r_d: loss resistance in the switching cell (semiconductors, connections, input capacitor);

– I_o: current in the load.

When the transistor is in open state:

$$I_k \approx e^{-\frac{t}{\tau}} \cdot I_o \cdot (\cos(\omega_1 t) + m \cdot \sin(\omega_1 t))$$

with

$$\omega_1 = \frac{1}{\sqrt{\ell_t C_t}} \cdot \sqrt{1-m^2} \ , \ m = \frac{r_d}{2} \cdot \sqrt{\frac{C_t}{\ell_t}} \ \text{ and } \ \tau = \frac{1}{m\omega_o}.$$

$$V_k = E + e^{-\frac{t}{\tau}} \cdot I_o \cdot \sqrt{\frac{\ell_t}{C_t}} \cdot \sin(\omega_1 t)$$

We note a first resonance state due to C_t and ℓ_t, and the static parameters (E, I_o) intervene over the amplitude of the oscillating elements. We need to minimize ℓ_t, which reduces the over-voltage over V_k and increases the damping.

As soon as the switch is in on-state, the current is initially in linear variation up until it reaches $I_0 + I_{RM}$; the diode is then in open state, followed by an oscillating state defined by[1]:

$$V_d = E' + e^{-\frac{t}{\tau}} \cdot \left(-E'\cos(\omega_2 t) + \frac{I_{RM} - m^2 \frac{2E'}{r_d}}{1-m^2} \cdot \sqrt{\frac{\ell_t}{C_d}} \cdot \sin(\omega_2 t) \right)$$

$$V_d \approx E + e^{-\frac{t}{\tau}} \cdot \left(-E\cos(\omega_2 t) + \frac{I_{RM}}{1-m^2} \cdot \sqrt{\frac{\ell_t}{C_d}} \cdot \sin(\omega_2 t) \right)$$

$$I_d \approx e^{-\frac{t}{\tau}} \left(I_{RM} \cos(\omega_2 t) + E\sqrt{\frac{C_d}{\ell_t}} \sin(\omega_2 t) \right)$$

1 This calculation assumes an instantaneous recovery of the diode (without losses), which overestimates the parasitic oscillation state.

with $E' = E - r_{dson} I_0 \approx E$, $\omega_2 = \frac{1}{\sqrt{\ell_t C_d}} \cdot \sqrt{1-m^2}$, $m = \frac{r_d}{2} \cdot \sqrt{\frac{C_d}{\ell_t}}$ and

$\tau = \frac{1}{m\omega_o}$.

The second resonance state appears when the diode is in open state, and C_d and ℓ_t must intervene. We note that this state is dependent on the inverse recovery current I_{RM}, which constitutes a major source of perturbation, by virtue of its amplitude and the speed of this phenomenon.

EXAMPLE 2.1.– ℓ_t = 450 nH, C_t = 1,500 pF and C_d = 500 pF, we obtain f_1 = 6.2 MHz and f_2 = 10 MHz. The spectrum of the parasitic signal shows peaks around these two frequencies. Figure 2.9 shows how these parasitic states modify the spectrum of the current inside the switching cell.

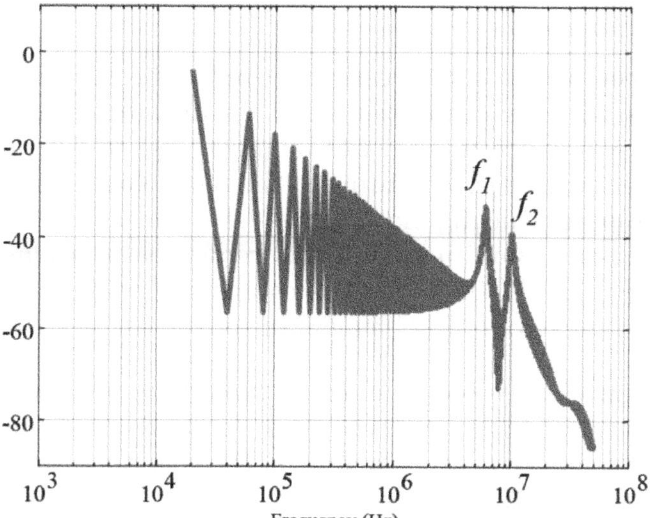

Figure 2.9. *Calculated spectrum of the current I_k taking the parasitic components into account, (unit dBA)*

These frequencies are on the scale of tens of megahertz. These signals are, therefore, very easily coupled, not only in a conducted mode by the parasitic capacitances of the structure, but also in radiated mode.

2.3. Characterization of coupling functions and parasitic elements

Coupling functions transmit a fraction of the energy converted toward the exterior of the converter in the form of undesirable parasitic common- and differential-mode currents. These functions depend on "physical" components in the structure and also on parasitic components that are "invisible" but which are preferably known and characterized if we wish to best control their perturbing effects.

2.3.1. *Passive components and differential-mode effects*

2.3.1.1. *Capacitors*

This component has a resistance and a parasitic inductance in series with its connections and its armatures to which a parallel resistance R_p is possibly added due to current leaks and dielectric losses. According to the technology, this element is more or less negligible. Figure 2.10 shows an impedance measurement as a function of the frequency for a plastic film capacitor of 1 µF/400 V and the comparison with an resistive inductive capacitive (RLC) model in series.

We can clearly see that in certain frequency ranges the different parameters of the model intervene immediately. The most inconvenient factor is the parasitic inductance which triggers the resonance of the capacitor at a specific frequency (~700 kHz in this example) followed by an increase in its impedance. Initially this is due to the connections of the component and its packaging.

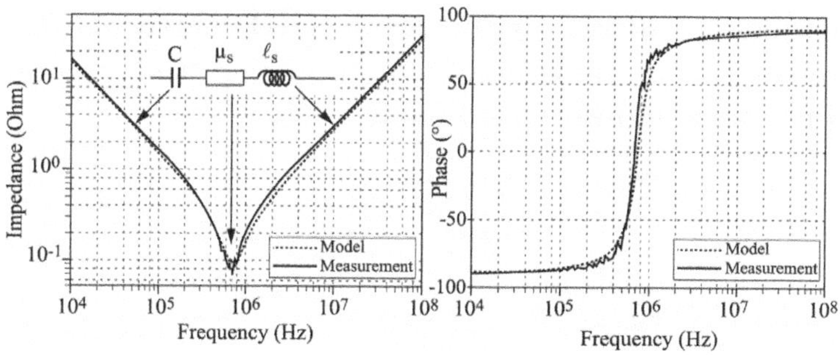

Figure 2.10. *Real impedance of a capacitor of 1 μF: $r_s = 85\ m\Omega$, $\ell_t = 43\ nH$, $C = 1.05\ \mu F$*

We have seen previously that the transmission of perturbations in differential mode depends on the common impedance that constitutes the input filtering capacitor. Thus, we can reduce the simplified model mentioned above to just differential mode as shown in Figure 2.11.

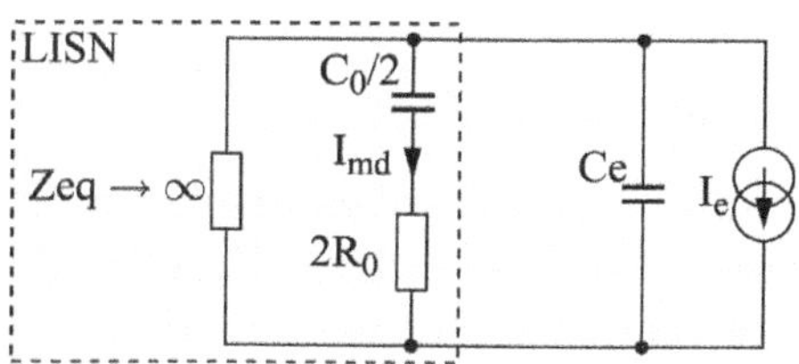

Figure 2.11. *Equivalent simplified model of the switching cell in differential mode*

The new expression of the current in differential mode, taking the parasitic effects of the capacitor C_e into account, will be:

$$I_{md} = \frac{C_0/2}{C_0/2 + C_e} \cdot \frac{1 + p^2 \cdot \ell_s C_e}{1 + 2R_o\left(\frac{C_e \cdot C_0/2}{C_e + C_0/2}\right) + p^2 \cdot \ell_s\left(\frac{C_e \cdot C_0/2}{C_e + C_0/2}\right)} I_e$$

The Bode diagram of the transfer function I_{md}/I_e is shown in Figure 2.12. We can see that the presence of the parasitic inductance ℓ_s increases the level of HF perturbations. Its minimization is imperative for reducing the amount of perturbations in differential mode.

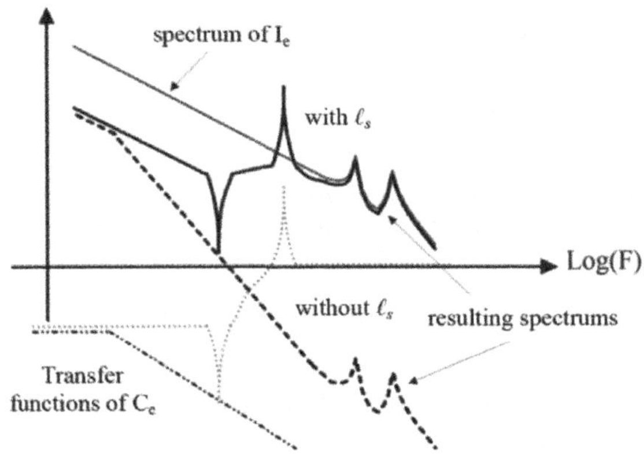

Figure 2.12. *Effect of the parasitic inductance of the input capacitor*

2.3.1.2. *Inductances*

In similar fashion, inductances also show imperfections at high frequencies. To illustrate, Figure 2.13 shows the measurement of the impedance module of an iron-core inductance used in a chopper. We can actually note that the inductive behavior is limited to approximately 100 kHz, and that beyond, the behavior becomes capacitive until approximately 2 MHz. This is followed by the apparition of resonances and antiresonances that confirm the propagating effects in the winding of the component. According to the frequency range on which we want to model the component, we are thus led to elaborate on its corresponding model by using an increasing number of cells r_i, l_i, c_i on a set scale. It is very difficult to quantify the value of these parameters using the geometry of the component and the physical properties of its constituents (magnetic materials, conductors and insulating materials). Their identification requires digital techniques of parameter optimization.

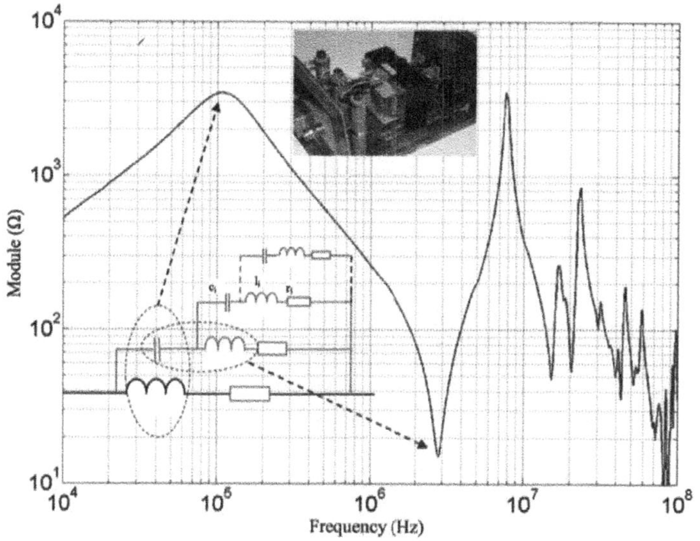

Figure 2.13. *Impedance of an iron-core coil and the effect from resistive, inductive and capacitive imperfections*

2.3.2. *Invisible parasitic elements and common-mode effects*

As we have already seen, power electronics components incorporate parasitic elements, which modify operation depending on the frequency. They are also the source of parasitic cross-talk (for example the radiation of magnetic components) or common-mode couplings. As an example Figure 2.14 shows the measurement of the common-mode impedance module of the iron-core inductance seen in Figure 2.13. The measurement was carried out between the input terminals in short circuit and the mechanical fastening point with the casing (itself connected to the earth potential). We observed a dominant capacitive effect up to 3 MHz, and beyond that, the propagation effects caused by the winding. Thus, at low frequencies, the model corresponding to the common-mode impedance in this example is simply a capacitance of 354 pF.

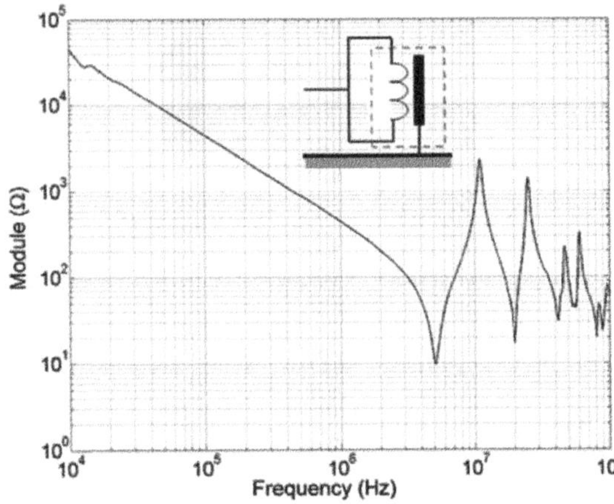

Figure 2.14. *Common-mode impedance module of an iron-core coil, $C_{p\_LF} = 354$ pF*

The common-mode couplings are very common in power-electronic structures, and exist virtually everywhere:

– At the power semiconductor level: simple components are insulated and mounted on a usually earthed heat sink. Therefore, there is a large parasitic capacitance between the drain (or collector), connected electrically to the metallic material of the casing, and the earth. The problem is identical for semiconductor power modules with regard to the base of the casing. We must note that this structural capacitance is connected to the middle point of the switching cell and is therefore also a structural common-mode current-injecting component, which depending on the technology, can be very contributory.

– At the connection level: printed conductors or cables have capacitances relative to the earth; even more so if they are shielded.

– At the load level: just like in the example seen previously, loads in power electronics (for example motors) have very significant common-mode impedances, which produce very large parasitic earth currents with extended spectrums as they are generally impulsive. Figure 2.15 shows the common-mode impedance module (the space between the framework and the three phases is short circuited) of a three-phase asynchronous motor of 400 V/3 kW: we observe a very capacitive behavior of the order of 6 nF at low frequencies (F < 50 kHz). This value is high and enables an easy transmission of parasitic currents in the earth connections of the machine, most notably when it is powered by a PWM inverter.

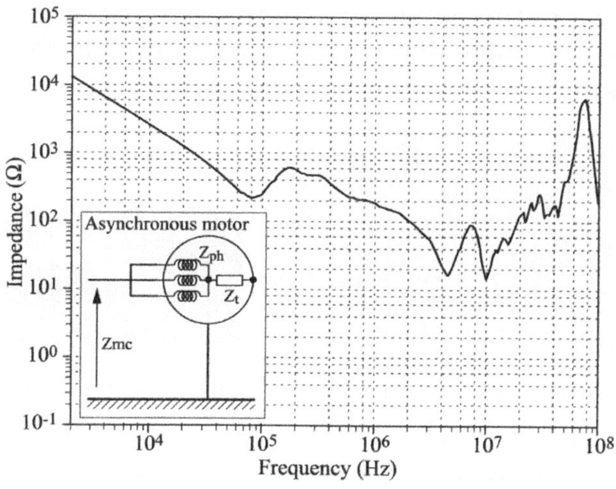

Figure 2.15. *Common-mode impedance module of an asynchronous three-phase motor 400 V / 3 kW*

2.3.3. *Parasitic effects contributing to undesirable couplings*

In power electronics devices, the connection between the components is generally ensured by printed circuits. In large power components or in the modules, the link between the exterior connections (pins, terminals, etc.) is ensured through the use of aluminum wires known as "bonding". All of these interconnecting elements introduce parasitic effects, which contribute to the transmission or creation of EM perturbations.

2.3.3.1. *Bonding*

Bonding wires (see Figure 2.16) are a heavily used interconnection technology in power electronics modules. This technology is well established and has the advantage of being sturdy and economical. Furthermore, it is tolerant toward the expansion caused by strong thermal constraints endured by the modules. The use of very fine materials (direct copper bonding (DCB) technology) in power modules greatly reduces the inductance of the electric tracks. The inductances of the connections to the exterior connectors and the parasitic inductances introduced through the bonding wires begin to dominate.

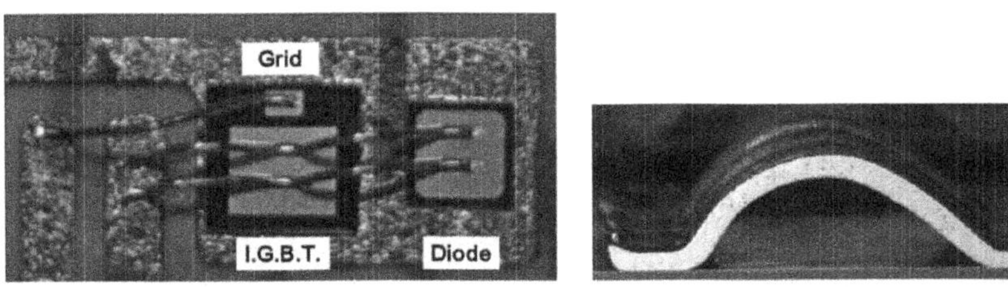

Figure 2.16. *Example of a bonding connection, cross-section view*

It is necessary to be able to determine the inductances introduced through bonding wires for the modeling of the parasitic inductances in the power electronics devices. On the one hand, the local effects, linked to the couplings between the bonding wires positioned parallel to each other, lead to a change in the distribution of the current in the MOS or IGBT transistor cells. This imbalance can also exist between parallel chips within the same module. The analysis of these imbalances can be facilitated through the analytical modeling of the couplings between the bonding wires, and linking it with the model of the track impedances of the module. On the other hand, the bonding wires used for the grid drive can be very close to the power bonding wires, and it is possible to model the couplings between these connections.

The diameter of a wire can vary between 100 and 500 m; the number of wires in parallel depends not only on the size of the current in the semiconductor, but also on the technical choice of the manufacturer of the module. The profile of the bonding wire used in the model consists of the height *h* of the wire's arc and the spacing *l* at the electric track level (Figure 2.17). By using 3D EM field calculation software, it is then possible to determine the magnetic field created around the bondings (Figure 2.18) and to identify their self and mutual inductances. As an example, a complete 3D model for a set of three aluminum bonding wires on an insulated magnetic substrate was created. The base is also made of aluminum (AlN: εr = 8.4). The electrical tracks are made of copper and have a thickness of 100 µm. The small thickness of the insulator greatly diminishes the inductance of microstrip tracks to which the bonding wires are soldered (their contribution in this case is virtually negligible).

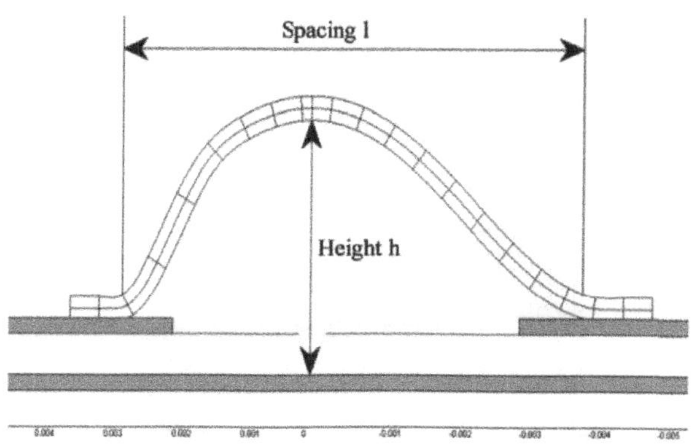

Figure 2.17. *Geometric representation of a bonding wire*

Figure 2.19 simultaneously shows the evolution of the potential and the distribution of the magnetic inductance field in the plane perpendicular to the wires. To precisely characterize the inductance introduced by the bonding wires, they have been placed between two microstrip tracks of identical size, whose linear characteristics are well known.

The contribution of these tracks to the total inductance can also be precisely calculated. This simulation configuration enables the comparison between the digitally obtained results and the measured results within the same context.

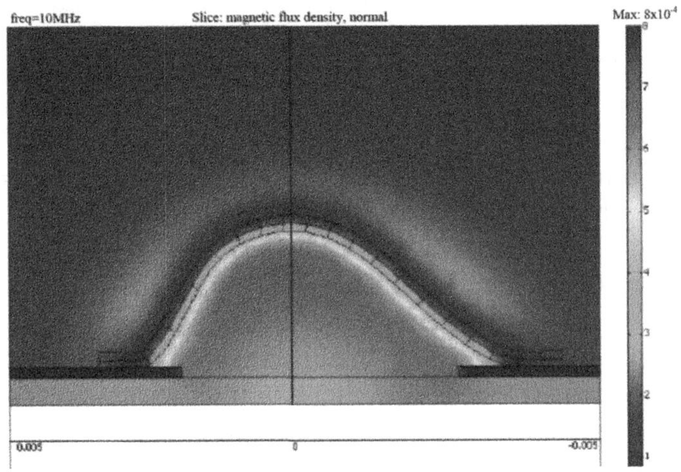

Figure 2.18. *Distribution of the magnetic field created by a bonding wire, F = 10 MHz. For a color version of this figure, see www.iste.co.uk / costa / electronics.zip*

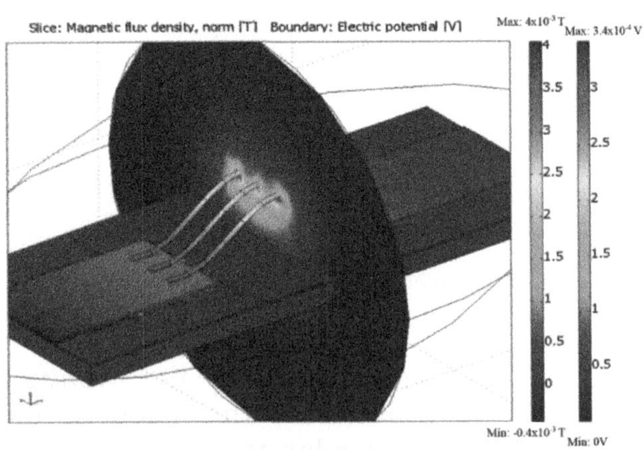

Figure 2.19. *Distribution of the magnetic field and the potential in the bonding wires, F = 10 MHz. For a color version of this figure, see www.iste.co.uk / costa / electronics.zip*

Table 2.1 shows the obtained inductance values for two types of substrate: insulated metallic substrate (IMS) and DBC.

	Global inductance	Inductance of the bonding wires
DCB 500 µm	5.6 nH	3.8 nH
IMS 100 µm	3.6 nH	3.0 nH

Table 2.1. *Calculation of the total inductance for three bonding wires in parallel*

Nevertheless, it can be interesting to take an analytical approach. The analytical model is based on the calculation of the inductance of a circular torus [SCH 00]. It is possible to determine the exact analytical expression of the inductance as a function of the geometric parameters of the torus: its average radius RT and the diameter of its section a (Figure 2.20). To fix the equivalent radius of the torus, the following two choices are possible:

– Equivalent radius R_{T1}: the equivalent length of the semi-torus is calculated relative to the real length of the bonding wire between the connections.

– Equivalent radius R_{T2}: the internal surface of the torus is equivalent to the real surface delimited by the bonding wire and the mass (Figure 2.17).

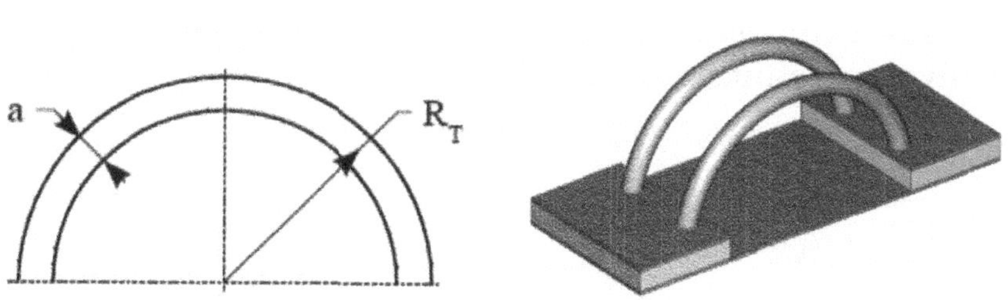

Figure 2.20. *Geometric representation for the analytical model*

The analytical formulas are provided below. They are given for a semi-torus above a mass plane. They make use of the first- and the second-degree integrals, K(k) and E(k), respectively. The term k represents the elliptic module of the integral.

$$L = \frac{\mu_0}{2}\sqrt{\frac{R_t + a}{R_t}}(2R_t - a)\left[\left(1 - \frac{k^2}{2}\right)K(k) - E(k)\right]$$

$$k^2 = \frac{4R_t(R_t - a)}{(2R_t - a)^2}$$

It is also possible to determine the expression for the mutuality between two wires, and the expression is:

$$M_{ij} = \frac{\mu_0}{2}\sqrt{\frac{(R_{ti} + R_{tj})^2}{D_{ij}^2} + 1}\left[\left(1 - \frac{k_{ij}^{\ 2}}{2}\right)K(k_{ij}) - E(k_{ij})\right]$$

$$k_{ij}^2 = \frac{4R_{ti}R_{tj}}{(R_{ti} + R_{tj})^2 + D_{ij}^2}$$

These expressions then allow us to determine the inductance matrix of the bonding connections for one or more semiconductors.

$$L_{matrix} = \begin{pmatrix} L_{11} & \cdots & M_{ij} \\ \vdots & \ddots & \vdots \\ M_{ij} & \cdots & L_{NN} \end{pmatrix}$$

To compare this with the example calculated by the 3D field software, Table 2.2 presents the L matrices as well as the global inductances obtained for the different choices of equivalent radiuses and substrates.

$$L_{1DCB} = \begin{pmatrix} 7,3e^{-9} & 2,7e^{-9} & 1,42e^{-9} \\ 2,7e^{-9} & 7,3e^{-9} & 2,7e^{-9} \\ 1,42e^{-9} & 2,7e^{-9} & 7,3e^{-9} \end{pmatrix}$$

$$L_{1DCB} = \begin{pmatrix} 6,5e^{-9} & 2,3e^{-9} & 1,2e^{-9} \\ 2,3e^{-9} & 6,5e^{-9} & 2,3e^{-9} \\ 1,2e^{-9} & 2,3e^{-9} & 6,5e^{-9} \end{pmatrix}$$

DCB 500 µm	R_{T1} = 3.6 mm	R_{T2} = 3.3 mm
Inductance	3.9 nH	3.4 nH
IMS 100 µm	R_{T1} = 3.2 mm	R_{T2} = 2.8 mm
Inductance	3.3 nH	2.7 nH

Table 2.2. *Results from the analytical calculations for the inductance matrices and for the global inductances of the three bonding wires in parallel according to different equivalent radiuses*

We can note that depending on the type of substrate, one or the other of the two choices of radius is closer to the result through 3D digital calculation (see Table 2.1); however in each case, an analytical approximation of 10% is permitted.

2.3.3.2. *The printed circuit*

Printed circuits contribute toward the generation and couplings of EM perturbations, even if their impedance is weak, and taking into account the incurred dynamic constraints (dV/dt, dI/dt), the resulting parasitic voltages and currents become non-negligible. Different methods exist, which are adapted to the frequency range operated by the circuit. We will only elaborate on one method that is well adapted to the frequency range of power electronics (DC to a few 10 MHz). It is the thin-wire method, which is, in fact, based on the Telegraphist's equations [CYR 01].

The thin-wire method is detailed in [PER 05], and it allows us to determine, from physical and geometric data, the behavior within a given frequency of a rectangular printed-circuit segment, which is "longer than it is wide". This method helps to quickly and precisely calculate the impedance of a printed-circuit segment while taking into account resistive, inductive and capacitive aspects. It also allows the calculation of couplings between different segments of parallel tracks. The initial method does not enable us to manage more complex configurations; the work we will present will demonstrate the validity of this method in a scenario with trapezoidal printed circuits and a scenario with couplings between non-parallel printed-circuit segments.

The complete model of the printed circuit requires the different tracks to be broken down into elementary segments. The electric characteristics of these segments, along with their couplings, are then calculated. To limit the complexity of the complete electric model, only the significant couplings are conserved in an initial optimization phase. The characteristics of the printed-circuit model can then be generated, with each segment being automatically given a directly interactive model in SPICE.

This principle is based on the equivalence between a microstrip line and a bundle of N cylindrical parallel wires between them (Figure 2.21).

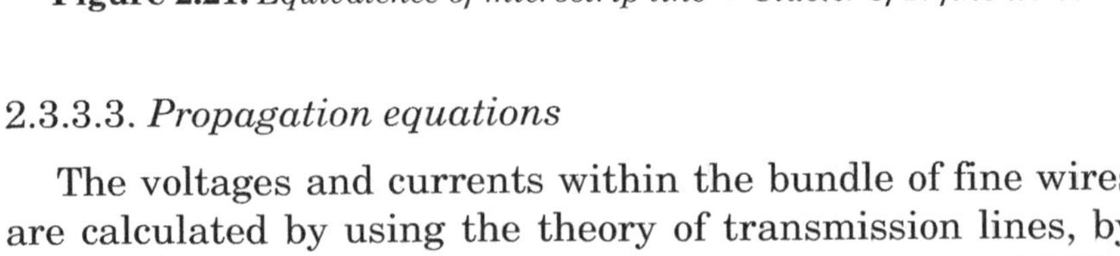

Figure 2.21. *Equivalence of microstrip line ⇔ Cluster of N fine wires*

2.3.3.3. *Propagation equations*

The voltages and currents within the bundle of fine wires are calculated by using the theory of transmission lines, by noting that the propagation is accomplished in quasi-TEM mode:

$$\frac{\partial[V(z)]}{\partial z} = -[Z_L]\cdot[I(z)]$$

where $[Z_L]$ is the linear impedance matrix of the cluster of thin wires

$$\frac{\partial[I(z)]}{\partial z}=-[Y_C]\cdot[V(z)]$$

where $[Y_C]$ is the linear admittance matrix.

The linear impedance matrix involves the resistance of conductors, the internal inductances, linked to the magnetic energy stocked in each conductor and the external inductances linked to magnetic fields outside the conductors. Due to the usual dimensions of the conductors that we use, we will ignore the internal inductance compared to the external inductance. In a sine operation with pulsation ω:

$$[Z_L]=[R(\omega)]+j\omega.[L_e]$$

The linear admittance matrix does not take into account the capacitances linked to the electric fields outside the conductors. Taking into account the frequencies for which we calculate these parameters, we suppose the dielectric is without losses. Let $[Y_C]$ be the linear admittance matrix with pulsation ω:

$$[Y_C]=j\omega.[C]$$

The literal solution to these voltages and currents for the cluster of fine wires is a linear combination of incident and scattered waves:

$$[I(z)]=e^{-[\gamma]z}.[I_+]+e^{+[\gamma]z}.[I_-] \quad \text{with } [\gamma]=\sqrt{[Z_L].[Y_C]}$$

$$[V(z)]=e^{-[\gamma]z}.[V_+]+e^{+[\gamma]z}.[V_-]$$

These equations allow us to calculate the currents and voltages at the input and output of the segment, depending on the frequency. They also allow us to calculate the couplings between the two segments of the printed circuit: we must, therefore, consider the set of thin wires necessary for the modeling of these two segments.

The external inductance matrix $\left[L_e\right]$ is determined by calculating the self-inductance of each conductor and the mutual inductances between every conductor (Figure 2.22). This calculation is based on the theory of partial inductances and helps to obtain an exact analytical expression in the case of cylindrical conductors with finite lengths and parallel to each other [SCH 94].

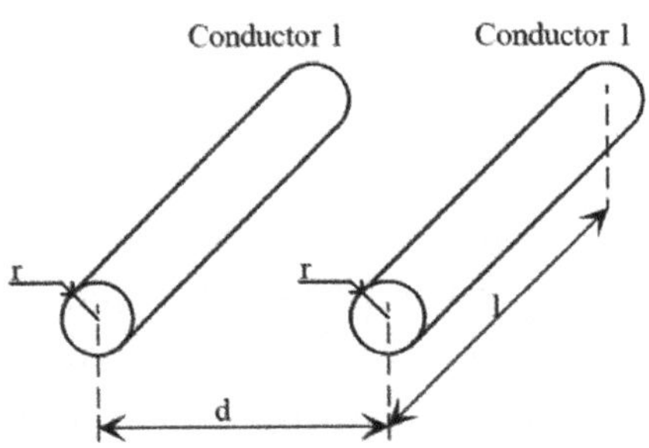

$$Lp_{ii} = \frac{\mu_0}{2\pi} l \left(\ln\left[\frac{l}{r} + \sqrt{\left(\frac{l}{r}\right)^2 + 1} \right] + \frac{r}{l} - \sqrt{\left(\frac{r}{l}\right)^2 + 1} \right)$$

$$Lp_{ij} = \frac{\mu_0}{2\pi} l \left(\ln\left[\frac{l}{d} + \sqrt{\left(\frac{l}{d}\right)^2 + 1} \right] + \frac{d}{l} - \sqrt{\left(\frac{d}{l}\right)^2 + 1} \right)$$

Figure 2.22. *Self-inductances and mutual inductances between two finite length cylindrical conductors*

The resistance matrix $\left[R(\omega)\right]$ is diagonal. The resistance of each conductor is calculated from the global resistance of a rectangular cross-section track, taking into account the variation in resistance depending on the frequency through the skin effect [CYR 01]. The expression used includes the Bessel and Kelvin functions.

By supposing a quasi-TEM propagation mode and introducing the notion of effective permittivity ε_{eff}, with the propagation medium not being homogenous, the relation between $[L_e]$ and $[C]$ is defined as such:

$$[L_e].[C] = \mu_0.\varepsilon_{eff}$$

This equation lets us determine the capacity matrix $[C]$ of the external inductance matrix $[L_e]$.

This method, therefore, allows us calculate the impedances of the printed circuits of a static converter and also evaluate its impact on conducted perturbations. An example is given in Figure 2.23(a), which shows a half-bridge inverter operated with a capacitive divider and MOSFET transistors, whereas its routing is shown in Figure 2.23(b). The printed circuit is placed 11.5 mm above the heat sink, which serves as a frame that drains the common-mode currents of the inverter. We have calculated all of the impedances of the structure's tracks as well as their common-mode capacitive couplings with the heat sink using the method previously described. They were then introduced into a SPICE-type "circuit" simulator as shown in Figure 2.23(c) as well as all of the other components of the inverter.

It is then possible to calculate the global parasitic current (common + differential mode) in the LISN and to determine its spectrum and then compare it with a direct measurement. Figure 2.24 shows that the spectrum is correctly predetermined up to 30 MHz.

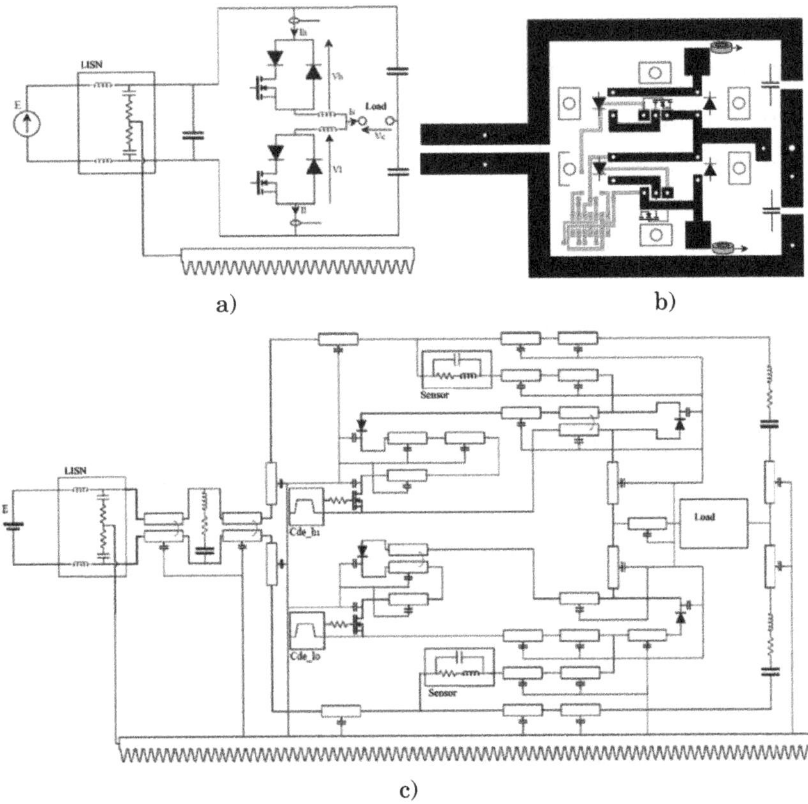

Figure 2.23. *a) Diagram of a single-phase inverter with a capacitive center point, b) routing, c) simulation diagram*

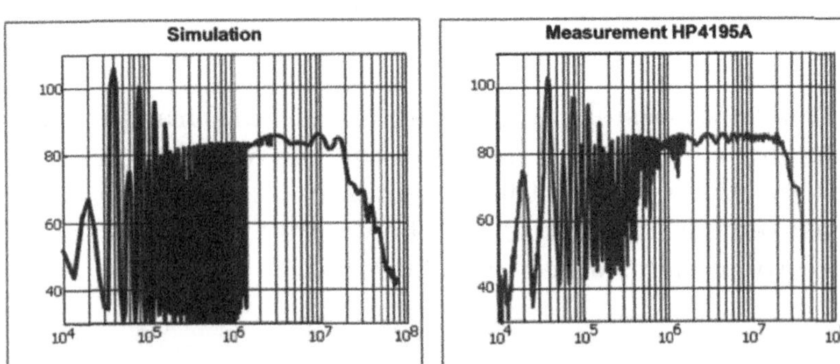

Figure 2.24. *Comparison of the calculated and measured spectra of the parasitic current in the LISN (unit dBμV, frequency in Hz)*

We can also quantify the impact of the printed circuit on the level of conducted perturbations in the LISN. We can then carry out the same simulation without introducing the printed-circuit impedances; in other words, the routing is ideal. We can, therefore, see in Figure 2.25 a very significant gap in HF in the 1–10 MHz range compared to Figure 2.24: the PCB contributes to the transmission of perturbations in this frequency range, most notably because of the common-mode capacitances that accompany it.

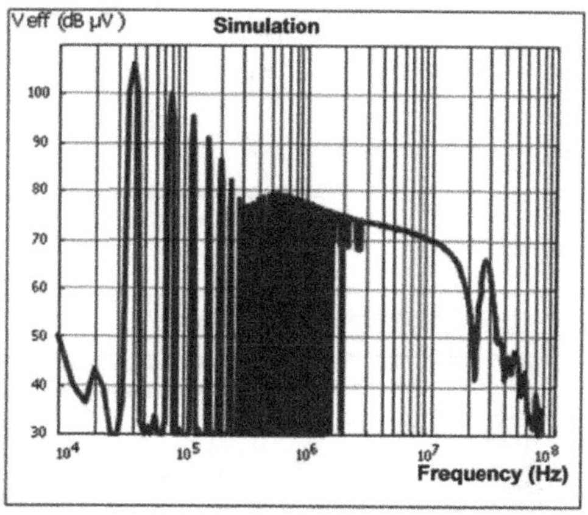

Figure 2.25. *Spectrum of the parasitic current in the LISN for an ideal printed circuit*

2.4. Electromagnetic compatibility study of a practical scenario: the Buck chopper

Following on from the electromagnetic compatibility (EMC) model of the switching cell previously presented (section 2.2), it is possible to illustrate the EM behavior of a converter with regard to the variations of the parameters involved in the sources or in coupling functions and to quantify their impact in terms of the standards. Nonetheless, a specific device must be studied that will allow us to obtain the variation in parameters, which is neither always possible nor observable in an industrial appliance.

2.4.1. *Description of the case study*

The case study is a Buck chopper, which is shown in Figure 2.26.

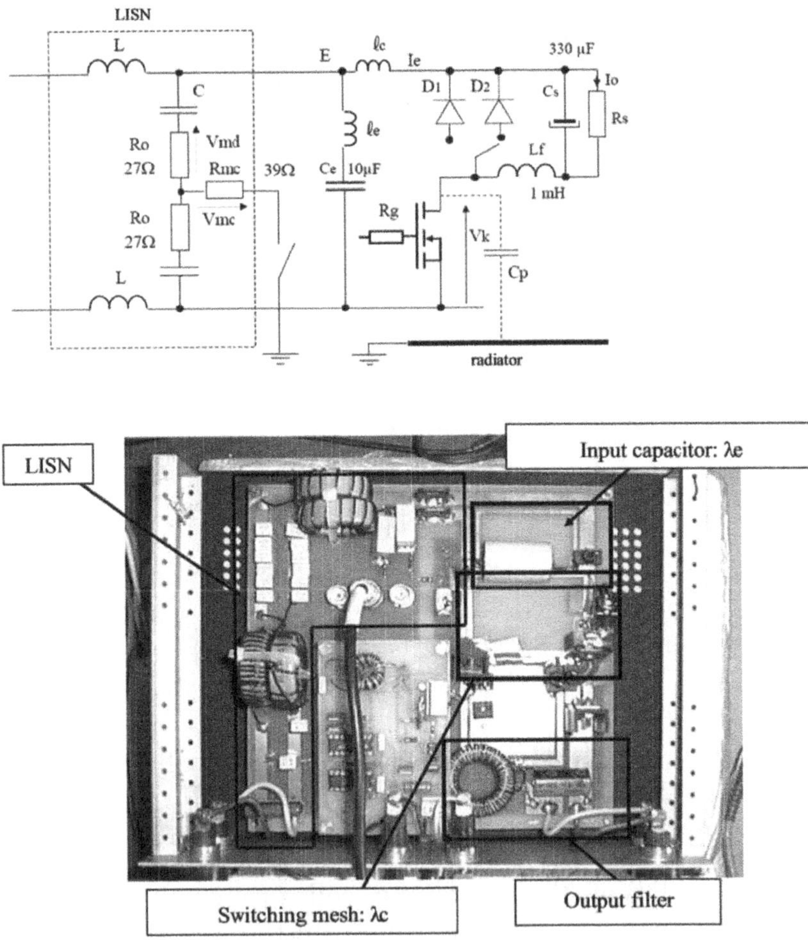

Figure 2.26. *Electrical diagram of the converter being studied and photo of the device*

This converter dedicated to the EMC study of a static converter was conceived to highlight the impact of the influencing parameters, linked to the following equivalent sources:

– the power supply voltage E: it can be up to 200 V at its maximum;

– the output current I_o: maximum 5 A;

– the technology of the semiconductors, notably the diodes, recovery of which is particularly disruptive: soft recovery for D1 and snap-off for D2;

– the switching speed of the MOSFET through action on the gate resistor R_g.

However, it is possible to modify the coupling functions:

– The inductance ℓ_c of the switching cell (see image) is adjustable between two values; this is obtained through a conductor printed onto the underside of the printed circuit, which is superimposed onto the switching cell. When this conductor is open, the value of ℓ_c is at its maximum, whereas when it is closed in short circuit, the fringing magnetic flux of the switching cell causes an inverse current in this conductor, which reduces the total created flow. As a result, the value of ℓ_c decreases.

– The inductance ℓ_e is also adjustable between two values; a loop is added in series with C_e, therefore artificially increasing the value of its parasitic inductance. It can be short circuited to obtain its minimum value.

– The value of C_p can be modified; in fact, two MOSFET transistors are wired and selectable in the switching cell, with one placed on a vertical heat sink with a weak C_p value, and the other is positioned directly onto a conductive large plane creating a high capacitance C_p with the ground plane of the converter.

Table 2.3 provides the values of the parasitic components, which are higher than in an industrial appliance, but, nonetheless, provide quantities on a similar scale.

	ℓ_e (nH)	ℓ_c (nH)	C_p (pF)
Maximum value	100	40	5
Minimum value	150	110	25

Table 2.3. *Values of the parasitic components of the chopper*

Finally, it is possible to separate the common- and differential-mode perturbations using LISN incorporated into the device; the common-mode current is observable (voltage *Vmc*) at the terminals of the resistance Rmc of 39 Ω. Thus, the common-mode impedance totals at 39 + 27/2 ~ 52.5Ω. We can remove the common-mode current and observe solely the differential-mode current at the terminals of a resistance R_o of 27 Ω by switching off the grounding of the LISN. The differential-mode impedance is 27 + 27 = 54 Ω. In the two propagation modes, we therefore have roughly identical impedances.

Figure 2.27 shows the differential- and common-mode perturbations spectrums of the chopper separately. The latter switches at F=15 kHz, is powered at 50 V, and its output current is of 2 A. We observe ray spectrums representative of the highly harmonic signals linked to the switching. These curves also let us illustrate the model previously outlined. The gray curve corresponds only to the differential-mode current circulating in the LISN; we can note that it is dominant in common mode (black curve) at low frequencies up to 300 kHz. We recall that the input current of the chopper is made up of a DC part to which an alternative quasi-triangular current is added due to the switching. Thus, in this frequency range, the switched current harmonics at the input are dominant. Their increase in frequency is roughly $1/f^2$: the equivalent source I_e of the differential-mode current (see Figure 2.1 and equations [2.2] and [2.4]) is a square waveform current, the spectrum of which decreases over *1/f*, linked to a coupling function $C_1(f)$ it also decreases over *1/f* in this frequency range. For a frequency of the order of 300 kHz, the envelope of the differential-mode spectrum passes through a minimum, which corresponds to the self-resonance of the capacitor C_e: at this frequency, the impedance of the capacitor reaches its minimum value (equal to its resistance in series), which minimizes the differential-mode current to be reinjected at the input.

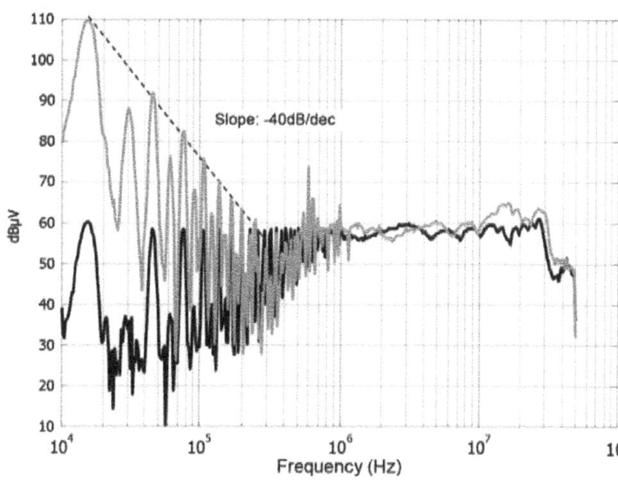

Figure 2.27. *Current spectrums in differential (gray) and common mode (black) measured in the LISN connected to the chopper: F = 15 kHz, E = 50 V, I_o = 2A*

We can confirm this particular frequency f_r through calculation by using the minimum value of ℓ_c in Table 2.3 associated with the value C_e = 10 µF:

$$f_r = \frac{1}{2\pi\sqrt{\ell_c Ce}} = 251\ kHz$$

Beyond this frequency, the amplitude of the parasitic differential-mode current spectrum is almost constant: this is explained by the fact that the decrease in the equivalent source I_e is always in $1/f$, but the function $C_1(f)$ is increasing in f starting from f_r (see Figure 2.5). Finally, in the 10–50 MHz range, we can see, on the one hand the existence of peaks due to resonances between the inductances and capacitances of the semiconductors when they are successively switched off, and on the other hand, we can see the decline of the spectrum because the source I_e is not a perfect square wave but it shows rise and fall times, which contribute to the decline of the spectrum in $1/f^2$. In the case of a perfectly trapezoidal signal, the increase time t_m and decrease time t_d of which are identical, the corresponding "cutoff" frequency f_c is given by:

$$f_c = \frac{1}{\pi t_m}$$

The amplitude of the common-mode current spectrum (black curve) is near-constant on a frequency range that stretches up until 20 MHz. This is explained by the fact that the source spectrum (voltage V_{mc}) decreases over $1/f$, whereas the coupling function $C_2(f)$ increases over f; indeed, common-mode coupling is essentially capacitive. According to equation [2.4], the result is therefore constant. Beyond 10 MHz, we find resonance peaks, the origin of which consists of the resonance from parasitic inductances of the switching cell with the capacitances of the switched off semiconductors. The final decrease in the spectrum has the same origin as previously mentioned.

These standardized measurements of the conducted perturbations do not differentiate between common and differential mode, with the previous analysis being only reliable for the analysis and comprehension of these phenomena. Figure 2.28 shows the conducted parasitic current spectrum (MD + MC) of the chopper measured by the

LISN. Thus, we also note the dominance of the equivalent differential-mode source I_e in LF. At high frequencies, it is the vectorial composition of the different harmonic orders of the common- and differential-mode currents which explain the appearance of the spectrum. Nonetheless, we deduce logically that the resulting spectrum MD+MC is always greater than the spectrum from either mode.

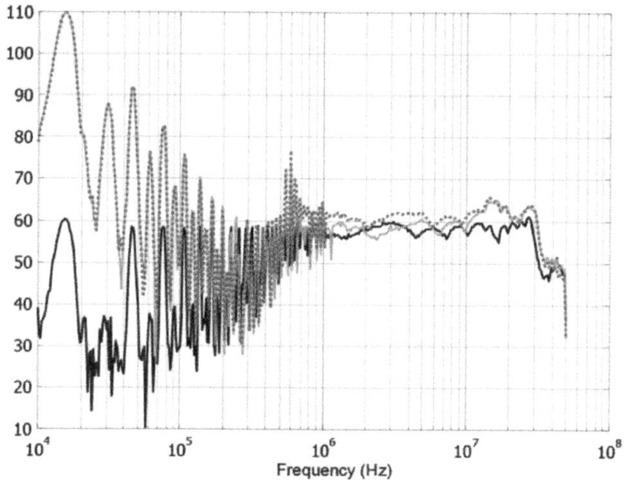

Figure 2.28. *Spectrum of the differential- (gray) and common-mode currents (black) along with the totals (dotted) measured in the LISN connected to the chopper: F = 15 kHz, E = 50 V, I_o = 2 A*

2.4.2. *Influence of the design parameters of the converter*

A static converter is designed for specific usage constraints: power-supply voltage, power to generate and operative frequency. Although the first two are effectively required for the usage of the device, the choice of switching frequency is usually free. The tendency is to increase the latter in order to reduce the dimensions of the converter and increase its adjustment frequency range.

Nonetheless, as we will note further on, the increase in switching frequency is evident through the increase in the level of perturbations. As a result, there is a necessary volumetric increase in filtering and shielding devices. Finally, a compromise between losses and EMC perturbations must be made to minimize the volume of the device.

Figure 2.29 shows the spectrums of the measured conducted perturbations in the LISN (superposition MC+MD) for two values of switching frequency f_d: 15 and 45 kHz. We observe a global increase in amplitude of the spectrum at high frequencies of the order of 10dB as a frequency ratio.

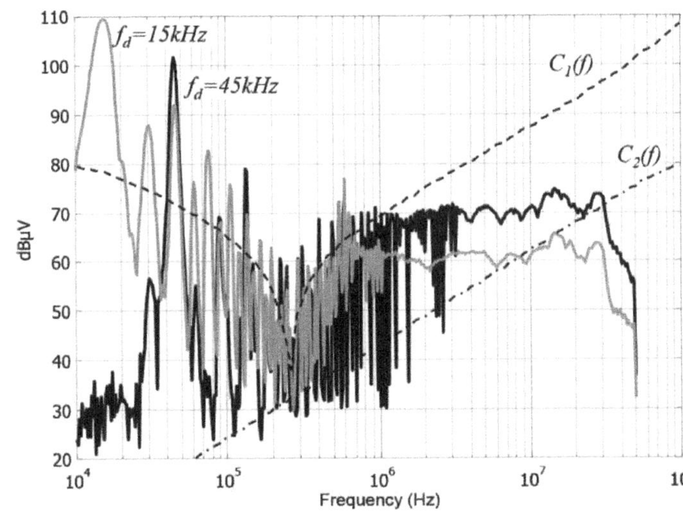

Figure 2.29. *Spectrum of the conducted currents (MC + MD) of the chopper for f_d = 15 kHz (gray) and f_d = 45 kHz (black). Dotted line: coupling functions $C_1(f)$ and $C_2(f)$*

This HF behavior is explained by the fact that the spectrums of the sources V_{mc} and I_e are shifted toward higher frequencies. Thus for a given harmonic order, the coupling functions remain unchanged (increase proportional to f), with the result being a gain in frequency ratios, of +10 dB. At LF, the reverse phenomenon applies: the function $C_1(f)$, dominant in LF, decreases over $1/f$. As a result, an LF harmonic order will be diminished at a factor of 3 or rather −10 dB, with this being particularly observable in the spectrum fundamentals.

It is equally important to note that the invariance of coupling functions implies that the HF resonances remain unchanged. Globally, in HF, we can see a transfer of the spectrum of perturbations to the ratio of the frequencies.

2.4.3. *Influence of technological parameters and control*

The technology behind power semiconductors and their control significantly determines the emitted perturbations. To illustrate, Figure 2.30 shows the evolution of the spectrum of the conducted perturbations of a chopper (F=15 kHz, V_{in} = 48 V, I_{out} = 2A) for two different diode technologies. We can see that when everything else is equal, the snap-off diode is clearly more perturbing than the soft diode.

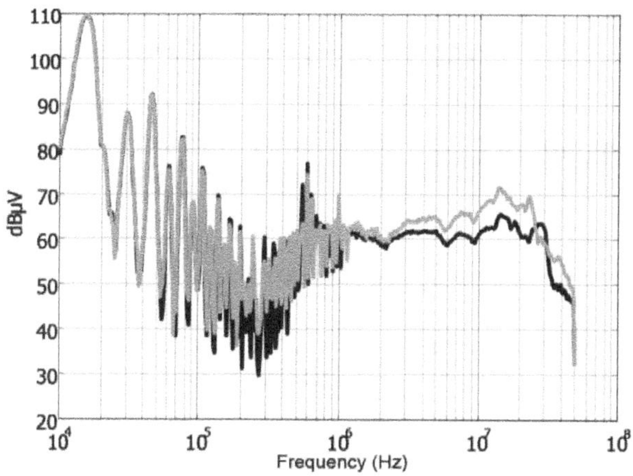

Figure 2.30. *Spectrum of a conducted perturbation (switching frequency F=15 kHz, V_{in} = 48 V, I_{out} = 2 A); gray: snap-off diode, black: soft diode*

Similarly, the gate control of power transistors (MOSFET or IGBT) influences the level of perturbation quite strongly, as shown in the curves of Figure 2.31, measured on the previous chopper. We observe considerable variations in the level of emitted noise at high frequencies (range of 1–10 MHz), as well as the excitation of intrinsic resonances in the conversion structure.

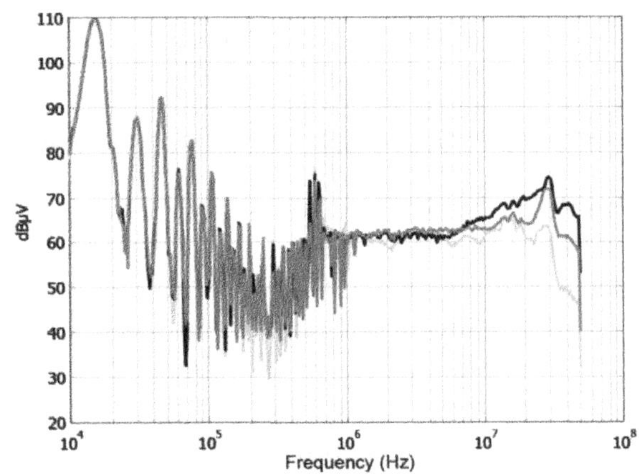

Figure 2.31. *Impact of the gate control of a MOSFET on conducted perturbations (F=15 kHz, V_{in} = 48 V, I_{out} = 2 A); light gray: R_g = 68 Ω, gray: R_g = 22 Ω, black: R_g = 0 Ω*

The consequences are obviously similar on perturbations radiated by the conversion structure.

2.4.4. *Other sources of switching noise*

Power semiconductors are controlled by field-programmable gate array (FPGA)-type digital circuits, microcontrollers or digital signal processors. The clocks of these devices are situated in a frequency range starting from a 100 MHz to a few gigahertz, and they are therefore also sources of significant EM noise, which frequently intervene beyond the perturbation frequency range of the power component. Figure 2.32 shows the spectrum of the field radiated by an inverter (working frequency: 4 kHz) powering an electric motor, with the high-amplitude rays (120–160 MHz) being a result of the badly shielded local quartz. We can see that in this frequency range (30–300 MHz), it is definitely the control that is perturbing and not the power stage (level tolerated "Class A, group 1–3 m").

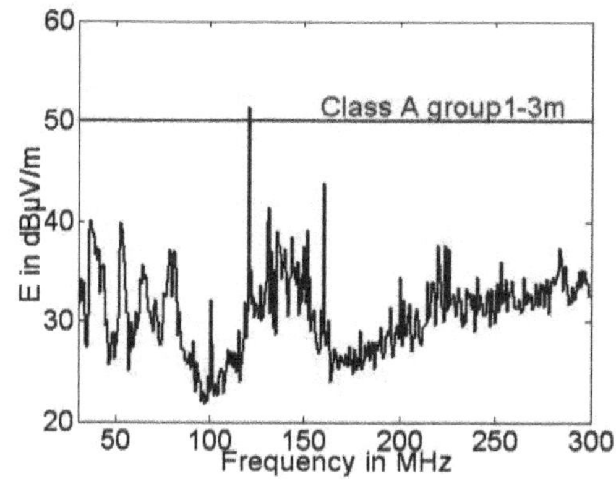

Figure 2.32. *Spectrum of the perturbations radiated by an inverter powering an asynchronous motor*

2.4.5. *Other switching modes: soft switching, advantages and constraints*

Through the introduction of reactive components into the switching cell, it is possible to "soften" the switching transition of voltages and currents and decrease their spectral envelope depending on the frequency quicker than with traditional switching control [SCH 00]. The switching cell is modified as shown in Figure 2.33. All of the quantities are equal, and we see the advantage of operating the converter in this mode.

The spectrum of the current deceases over $1/F^3$, inversely to the traditional switching scenario where it decreases initially over $1/F$ then over $1/F^2$. The result is, therefore, a weaker impact in terms of HF perturbations.

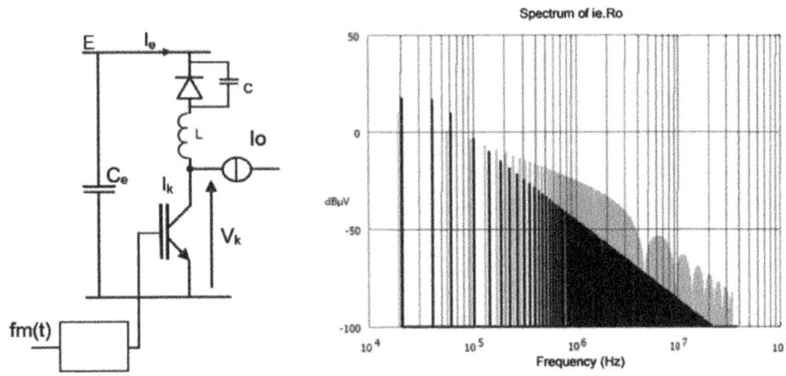

Figure 2.33. *Soft-switching conversion structure, comparative spectrums of the current I_e; black: soft switching, gray: hard switching*

2.5. EMC study of an insulated DC-DC flyback power supply

2.5.1. *Description of the device*

The switch-mode power supply being studied is part of an industrial variable speed drive and provides all of the necessary voltages to the drive: drivers, control board, etc. It has a power of 50 W, an input voltage of 550 V DC and its switching frequency is of 50 kHz. It is of the multi-secondary flyback type. Nonetheless, we have noted that in the majority of operating states of the variable-speed drive, one of the windings provides the majority of the power at a voltage of 16 V. We will, therefore, limit the study to the simplified case of one single secondary winding as shown in Figure 2.35. We have seen that it generates a significant level of EM perturbations, notably radiated perturbations. The conversion structure is a traditional one: the main switch, made up of the high-voltage bipolar transistor K_p and of the low-voltage MOSFET K_{aux} is of the cascode type, which enables quick switching with low conduction losses. The circuits D_1, C_1, R_1 and D_2, C_2, R_2, respectively, make up the passive clamp and snubber circuits of the structure.

Common- (I_{mc}) and differential-mode (I_{md}) currents circulate in the transformer, shown as the dotted line in Figure 2.35.

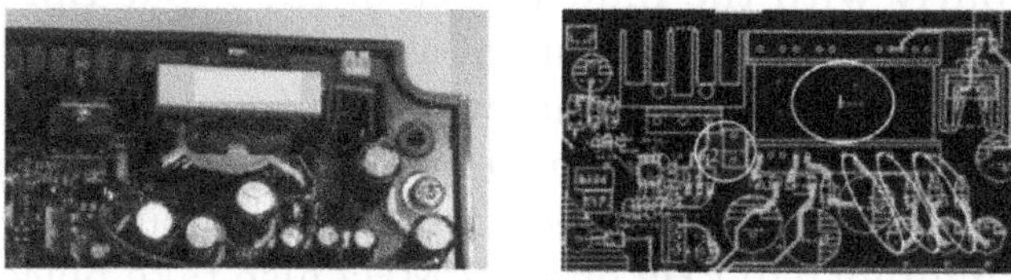

Figure 2.34. *View of the switch-mode power supply and the layout of its components*

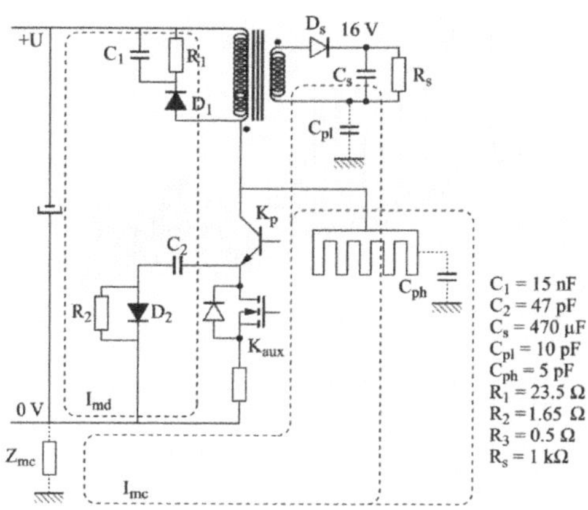

Figure 2.35. *Flyback switch-mode power supply*

Figure 2.34 shows a view of this power supply as well as its layout. We note its compactness and the very compact routing of the switching cell and snubber circuits (white ellipses). Thus, the magnetic near-field radiation is essentially due to the shape of the transformer, its physical properties and the currents that traverse it. Taking into account the topology of the converter (non-earthed heat sinks of semiconductors), the common-mode currents are of the second order due to the weak values of the parasitic common-mode capacitances C_{ph} and C_{pl}. We will, therefore, focus the study on to the dominant differential-mode current in the transformer.

The major drawback of the near-field radiation is the coupling of this parasitic field with sensitive components, particularly with the inductances of the input EMC filter of the hardware. A measurement was taken above the power supply using a loop antenna, with the frequency spectrum of the field H_z being shown in Figure 2.36. We observe a global decrease at −20 dB/dec where resonance peaks of approximately 540 kHz and 4.2 MHz also take place. We will later explain the origin of these phenomena.

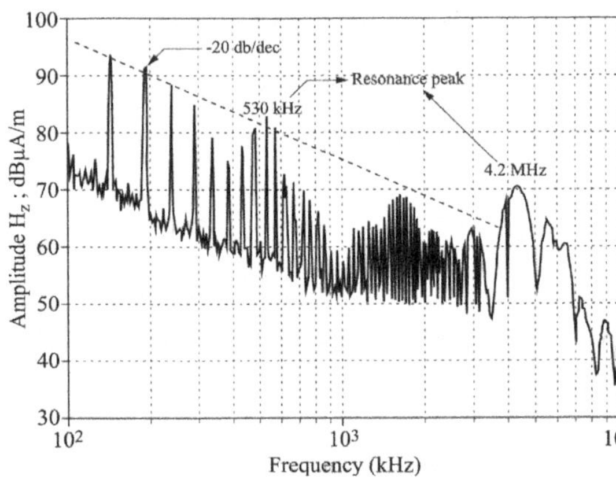

Figure 2.36. *Spectrum of the near magnetic field, measured 7 cm above the transformer of the power supply*

To refine this observation, we have also mapped out the magnetic H_z field being radiated above the power supply. It is shown in Figure 2.37 at 540 kHZ and at 4.2 MHz, frequencies which are highlighted in the spectrum in Figure 2.36. We take note of the maximum level of radiation centered above the transformer at both frequencies, which confirms its major status as a source. The affected zone is very large, and, in particular, touches the location of the input EMC filter of the variable-speed drive, which creates a parasitic coupling limiting its efficiency.

The methodology to be learned consists of determining the current in the transformer, which therefore leads to creating a circuit "model" of the structure, and secondly in establishing the relation between this current and the field radiated above the transformer. The latter depends on its shape and the physical properties of its components.

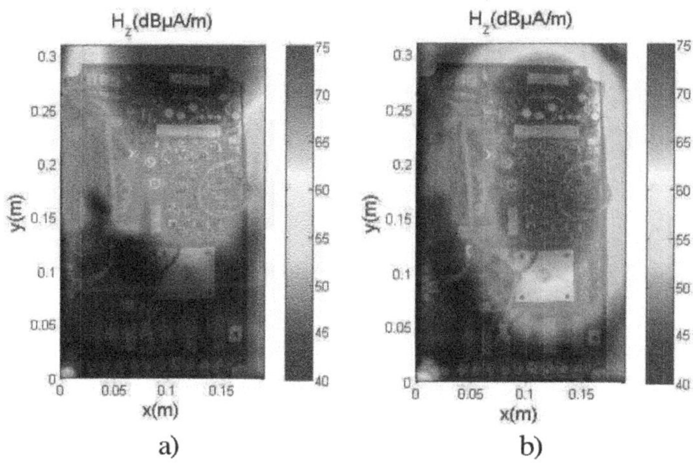

Figure 2.37. *Mapping of the hertz field at z = 10 cm above the PCB: a) 530 kHz and b) 4.2 MHz. For a color version of this figure, see www.iste.co.uk / costa / electronics.zip*

2.5.2. *Creation of the circuit model*

The modeling principle of the insulated structure is identical to that seen in section 2.2. It is, nonetheless, necessary to define the circuit model of the transformer. Following on from the equivalent traditional diagram in which we add the electrostatic effects created by the capacitances, as shown in Figure 2.38(a), we can outline two equivalent variations (b) and (c).

The terms R_s, L_f, R_f and L_μ, respectively, refer to the total resistive losses brought back to the primary, the total magnetic flux leaks brought back to the primary, the total magnetic losses, the total losses in ferrite material of the magnetic circuit and the magnetizing inductance. The capacitances C_{pp}, C_{ps}, C_p represent the electrostatic effects at the primary, at the secondary, and between the primary and the secondary, respectively.

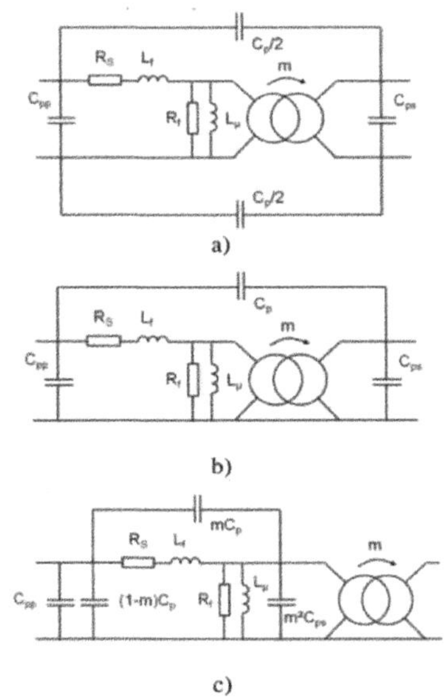

Figure 2.38. *Models of the transformer of the flyback power supply*

This "circuit" model can also be represented in the form of an "admittance" quadrupole the terms of which are defined below:

$$\begin{cases} i_1 = Y_{11} \cdot V_1 + Y_{12} \cdot V_2 \\ i_2 = Y_{21} \cdot V_1 + Y_{22} \cdot V_2 \end{cases}$$

With:

$$\begin{cases} Y_{11} = j\omega\left(C_p + C_{pp}\right) + \dfrac{1}{R_S + jL_f\omega} \\ Y_{22} = j\omega\left(C_p + C_{ps}\right) + \dfrac{1}{m^2}\left(\dfrac{1}{R_S + jL_f\omega} + \dfrac{1}{R_f} + \dfrac{1}{jL_\mu\omega}\right) \\ Y_{12} = Y_{21} = -j\omega C_p - \dfrac{1}{m}\left(\dfrac{1}{R_S + jL_f\omega}\right) \end{cases}$$

By measuring the input impedance of unloaded transformer, then at the secondary in short circuit and between the primary and secondary in short circuit, we can identify the values of the different parameters by observing the behavior at LF (the L and R terms dominate) and at HF (C terms dominate). To illustrate this, Table 2.4 gives the values obtained from the transformer.

Elements	***Values***
C_{pp}	20 pF
C_{ps}	1,200 pF
C_p	15 pF
L_μ	0.42 mH
L_f	28 µH
R_S	4 Ω
R_f	100 kΩ
M	0.19

Table 2.4. *Values of the elements in the equivalent model of the transformer*

Figure 2.39 shows the evolution of the input impedance of the transformer, with unloaded secondary, as a function of the frequency. We superimposed the curve obtained through simulation and the measured curve; the superimposition is correct up until 4.5 MHz. The equivalent model of Figure 2.38(c) allows us to easily interpret the evolution of the curve and its different resonances and antiresonances observed while the frequency increases. At low frequency, the terms R_S, L_f, R_f and L_μ determine its evolution. The first peak corresponds to the parallel resonant circuit created by C_{pp}, $(1 - m)C_p$ and m^2C_{ps} and L_μ; the first low point corresponds to the resonance in series between m^2C_{ps} and the leak inductance L_f: the second peak corresponds to the parallel resonance between L_f and mC_p connected in parallel with C_{pp} $(1 - m)C_p$ and mC_{ps} in series. Finally at HF, the capacitive terms become dominant. The amplitudes of the maximum and minimum values of the curve are different to those of the measured curve. This is due to not taking into account the evolution of R_S and R_f with the frequency (skin effect and proximity) as well as the losses linked with electrostatic couplings.

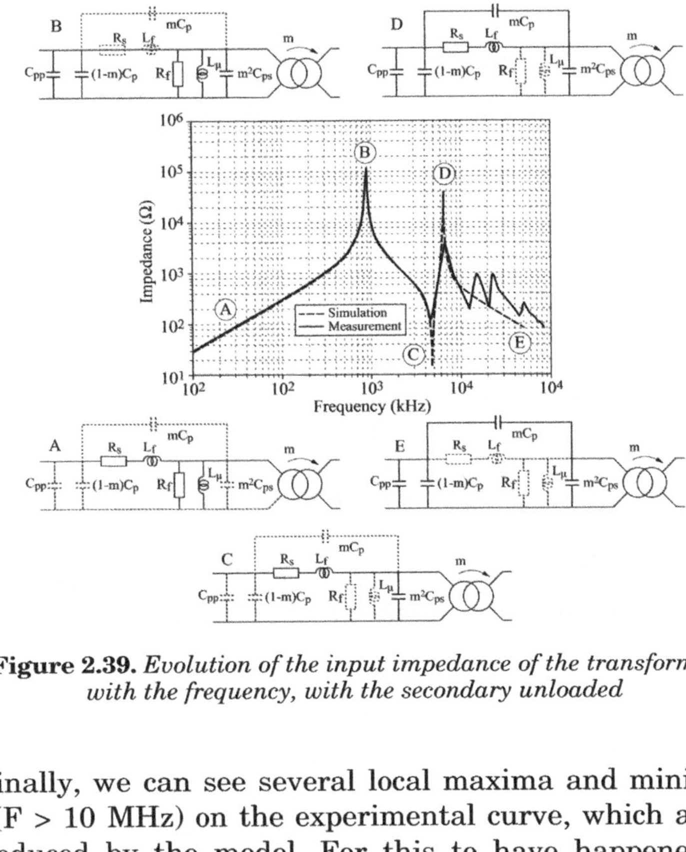

Figure 2.39. *Evolution of the input impedance of the transformer with the frequency, with the secondary unloaded*

Finally, we can see several local maxima and minima at HF (F > 10 MHz) on the experimental curve, which are not reproduced by the model. For this to have happened, the order of the model would have had to be increased by adding the corresponding R-L-C-m terms to the other windings of the transformer which, as we recall, were not considered for this simplified approach. Nevertheless, we can see a strong consistency in the frequency range up to 10 MHz, which is sufficient for this example.

2.5.3. *Analysis of switchings in the structure*

This structure is simple, but the presence of the snubber circuits complicates the switching phases of different semiconductors, which are frequently the source of barely dampened high-frequency oscillating states that contribute to the conducted and radiated perturbations. An initial analysis of the simplified electrical operation (without taking note of the electrostatic effects that generate oscillations) is, therefore, necessary to be able to analyze the curves obtained from the numerical simulation. This analysis is based on the electrical diagram in Figure 2.41.

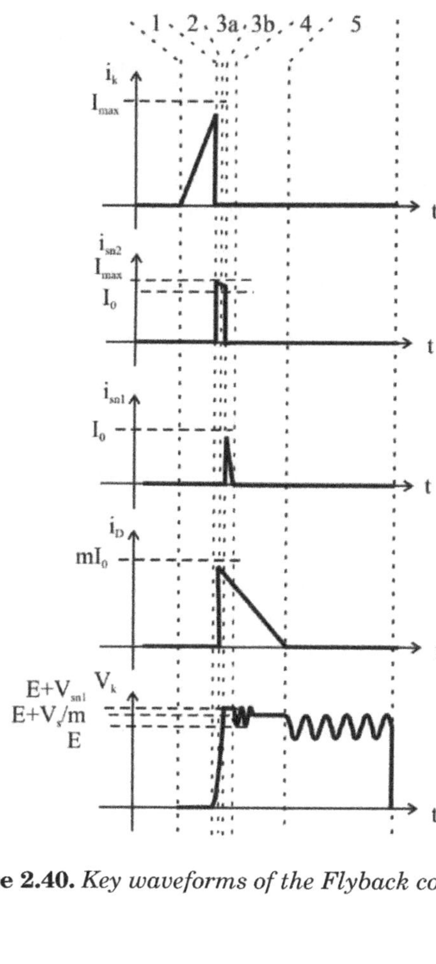

Figure 2.40. *Key waveforms of the Flyback converter*

Phase 1: the switch K is off. The other semiconductors are on. The current i_k increases linearly.

Phase 2: when we quickly turn off the switch K, the primary current i_k imposed by the primary inductance activates the diode D_{SN2} (snubber), C_K and C_2 charge in a short space of time. The voltage V_K at the terminals of C_k increases.

Phase 3a: when V_k is greater than E + V_S/m, (V_S/m is the output voltage brought back to the primary), the secondary diode D begins to activate.

Phase 3b: when the voltage V_k reaches E + V_{sn1}, the diode D_{SN1} activates and the limiting circuit RC_1D_{sn1} absorbs the energy still present in the leakage inductance L_f. The primary current i_t, therefore, travels through C_{SN1} via diode D_{SN1}. The voltage V_{sn1} at the terminals of the capacitor C_{SN1} gently increases. This phase is finished once the leaked energy is stored in L_f and entirely dissipated into the RCD circuit, in other words once i_{sn1} is at 0.

Phase 4: the diode D transfers the energy stored in L_μ to the secondary circuit, and V_K is equal to E + V_S/m. This phase is finished once i_D is at 0.

Phase 5: all of the switches are on, the voltage V_K returns to the value E through a dampened oscillating state.

Phases 2, 3a and 3b last a very short amount of time. If we take into account the presence of parasitic electrostatic effects (complete model of the transformer and the semiconductors), each commutation is accompanied by a resonance, its frequencies of which depend on the involved parasitic components.

2.5.4. *Electric simulation of the complete structure*

We have noted that the presence of the two assistance circuits (snubber circuits) add additional switching steps and consequently associated parasitic states due to the multiple parasitic components involved (transformer, semiconductors, cabling). This complex operation, therefore, must be simulated by a "circuit"- and SPICE-type calculation software. The power supply model is shown in Figure 2.41. For the sake of simplicity, we have replaced the switches and diodes with ideal switches, where we have simply added their supposedly constant parasitic components in parallel to the terminals: it is the only parasitic effect taken into consideration.

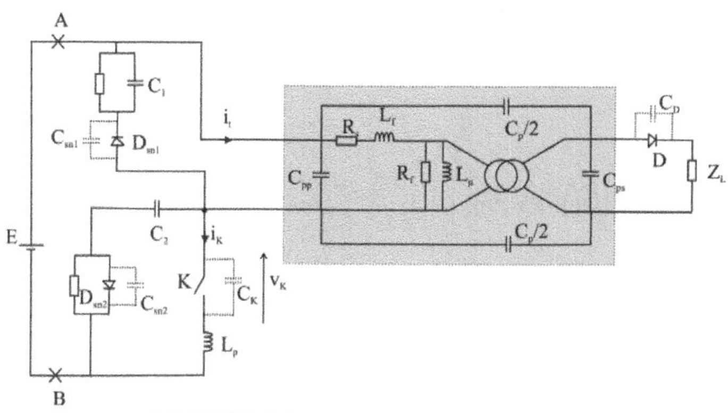

Elements	Values
C_{sn1}	60 pF
C_{sn2}	30 pF
C_k	40 pF
L_p	70 nH
C_1	15 nF
C_2	47 pF
C_D	50 pF

Figure 2.41. *Electrical diagram of the complete structure of the flyback and the values of the components*

These are big approximations but, as we will demonstrate, they allow us to obtain qualitatively acceptable results, which are quantitatively close enough to measurements. The load is represented by Z_L, which encompasses the resistance and output capacitor.

The voltage at the terminals of the primary switch is shown in Figure 2.42, and we can see that the oscillations during the cutoff of different semiconductors are correctly simulated (however, the damping of the main oscillation differs), which backs up the validity of the model when reproducing the dominant effects, particularly the major oscillation at 540 kHz during that cutoff of diode D. Similarly, we calculated the current i_k in the switch and the current i_t in the transformer and then calculated their frequency spectrum through FFT.

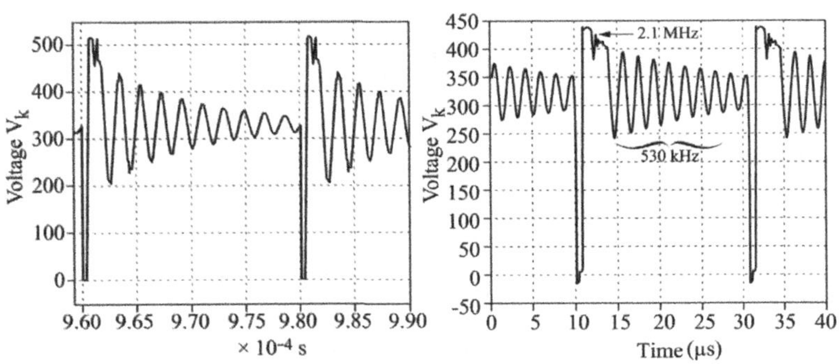

Figure 2.42. *Simulated voltage and voltage measured at the terminals of the primary switch*

We note a good consistency between the spectrums of the current i_k in Figure 2.43(a) and (b). If we compare the spectrums of field H in Figure 2.26 and the current i_t in Figure 2.43(c), we can see that in the frequency range from 1 to 10 MHz, the lobes coincide significantly enough to be able to correlate the current i_t calculated in the transformer, with the radiation from the power supply.

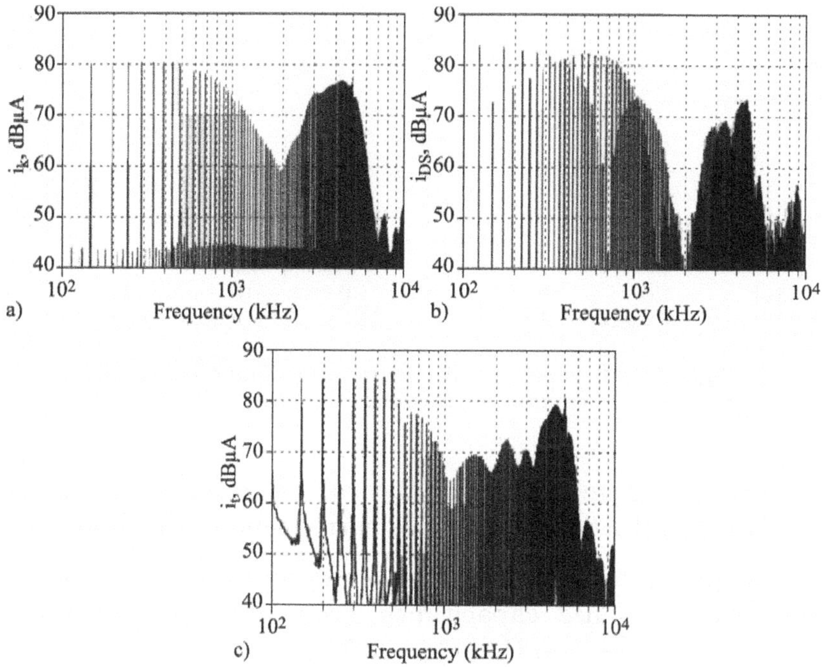

Figure 2.43. *Spectra of the current I_k in the switch: simulated a) and measured b); spectrum of the current I_t in the transformer: c) calculated*

The study of the corresponding "circuit" diagram also allows us to predetermine the resonance frequencies by considering the equivalent diagram in each commutation phase and by studying the impedance Z_{AB} found between the terminals A and B, which correspond to the input mesh of the converter (see Figure 2.41). A resonance from the mesh corresponds to a zero impedance. Thus, the analysis of the evolution of Z_{AB} during the different phases of operation lets us predict the resonances of i_t. Figure 2.44 shows the evolution of Z_{AB} for phases 1, 4 and 5. We find as expected a zero at 4.5 MHz for phase 1 (4.2 through measuring) and two zeroes for phase 5 at 490 kHz and 5 MHz (530 kHz through measuring).

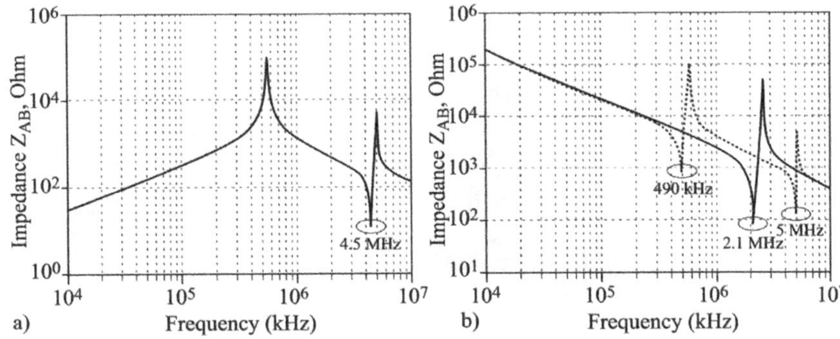

Figure 2.44. *Impedance Z_{AB} for a) phase 1 and b) phases 4 (continuous line) and 5 (dotted)*

Once we develop the analytical expressions for these impedances, the two frequencies correspond to the following:

– In phase 5: resonance between the inductance L_μ and the total parasitic capacitance seen by this inductance: $C_k + C_{sn2} + C_{sn1} + (1\text{-}m)C_p + C_{pp} + m^2C_{ps}$:

$$f = \frac{1}{2\pi\sqrt{L_\mu\left(C_k + C_{sn2} + C_{sn1} + C_{pp} + (1-m)C_p + m^2C_{ps}\right)}} = 520\text{ kHz}$$

– In phase 1: resonance between L_f and the parasitic capacitance of the transformer m^2C_{ps}:

$$f = \frac{1}{2\pi\sqrt{m^2L_fC_{ps}}} = 4.5\text{ MHz}$$

Thus, this simplified model lets us highlight the main conducted and radiated perturbation mechanisms in near-field to a flyback power supply. Evidently, it would be necessary to refine it in order to improve the precision of the simulation, but using the same methodology.

2.6. Corrected exercise number 1: conducted perturbations of a step-up chopper

Figure 2.45 shows the spectrum of high-frequency perturbations measured at the input of a sinusoidal BOOST (power factor corrector (PFC)), connected to the mains of 230 V/50 Hz. Its electrical and topological diagram is shown in Figure 2.46, with the switching frequency at 85 kHz. The measurement is carried out using an LISN, the corresponding diagram of which is given in Figure 2.47 and is simplified with regard to the propagation of high-frequency perturbations. We wish to estimate its perturbations in view of the class A EN55011 standard (see the graph below).

We suppose that the output voltage is Vs = 400 V, and we call $f_m(t)$ the modulation function of the chopper.

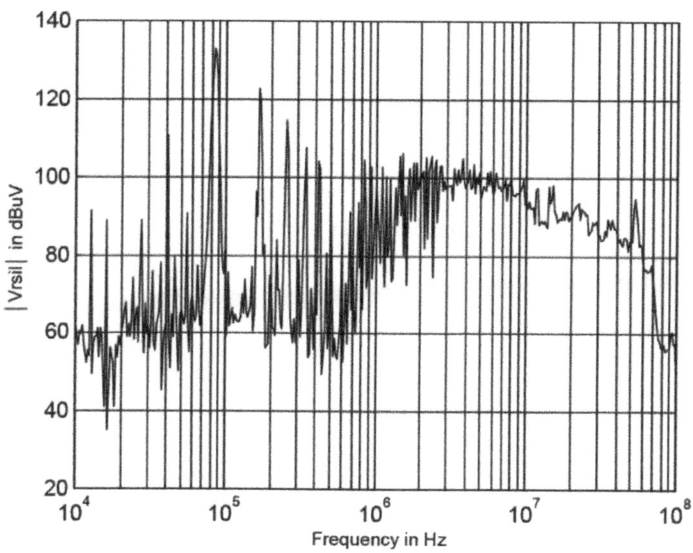

Figure 2.45. *Spectrum of the perturbations measured at the input of the converter at the terminals of the resistance R_{o1} of the LISN*

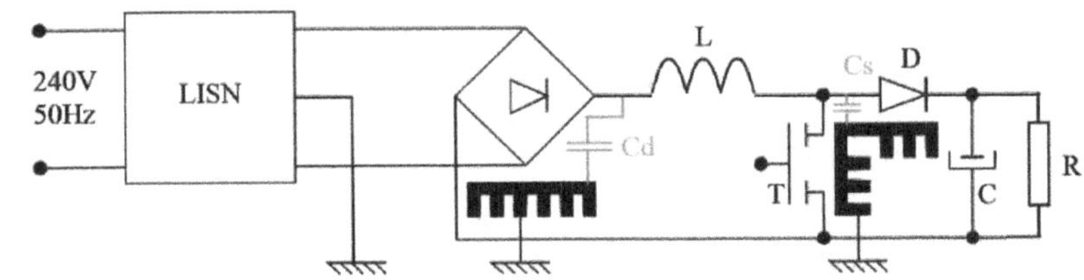

Figure 2.46. *Structure of the converter*

We characterize the inductance L using the network analyzer up to 30 MHz. The measurements have shown that a parallel RLC model is adequate. The values of the different elements are given here: L = 883 μH; r_p = 19 kΩ; C_p = 105 pF.

The parasitic capacitances (in gray in Figure 2.46) between the semiconductors and the switching stage (Cs) of the rectifier with input diodes (Cd) and the dissipater were measured, and are Cs = 25 pF and Cd = 30 pF.

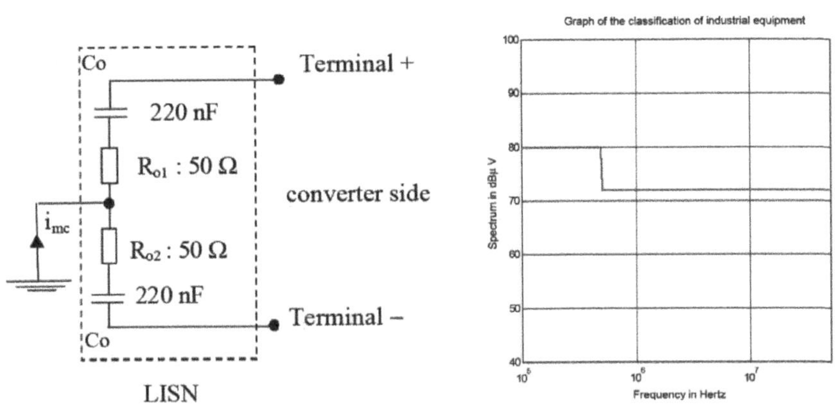

Figure 2.47. *Simplified equivalent HF diagram of the LISN and standards graph*

1) Give the expression Z_L of the inductance L taking into account its parasitic elements. From what frequency value does its behavior change?

2) Propose a complete equivalent EMC diagram (common and differential mode) of this converter, based on the EMC model of the switching cell, including the equivalent sources.

3) With a view to respecting the radiation standards (EN55011), which source is it acceptable not to take into account and why? We will adopt this simplification later on.

4) Define within the time and frequency ranges (variable *s*) the main equivalent source as a function of Vs and $f_m(t)$. Trace the approximate shape of its frequency spectrum on a semi-logarithmic scale and in dBμV (supposing that $f_m(t) = 0.5$).

5) Calculate the voltages at the terminals of the resistances R_{o1} and R_{o2} with:

$$V_{rsil1}(s) = C_1(s).\ f_m(s).Vs \text{ and } V_{rsil2}(s) = C_2(s).\ f_m(s).Vs$$

From this, deduce the expression of the common-mode current $i_{mc}(s)$ in the LISN.

6) Identify the coupling functions $C_1(s)$ and $C_2(s)$ according to the different elements of the structure.

7) The coupling functions $C_1(f)$ and $C_2(f)$ were traced in Figure 2.48. Interpret the shapes of the curves (local minima, maxima, HF asymptotes) with regard to the parasitic effects of the components of the structure.

8) Show that there exists an asymmetry between the values of the perturbations measured at the terminals of the two resistances R_{o1} and R_{o2}.

(The following two questions will be discussed in Chapter 4, section 4.6, regarding filtering).

9) Propose a qualitative interpretation of the envelope of the spectrum of perturbations in Figure 2.45 (decrease in the rays, minima, etc.) relative to the curves in Figure 2.48.

10) Does the actual converter abide by the EN55011 standard? If not, propose and justify a structure for an EMC filter adapted to this conversion structure.

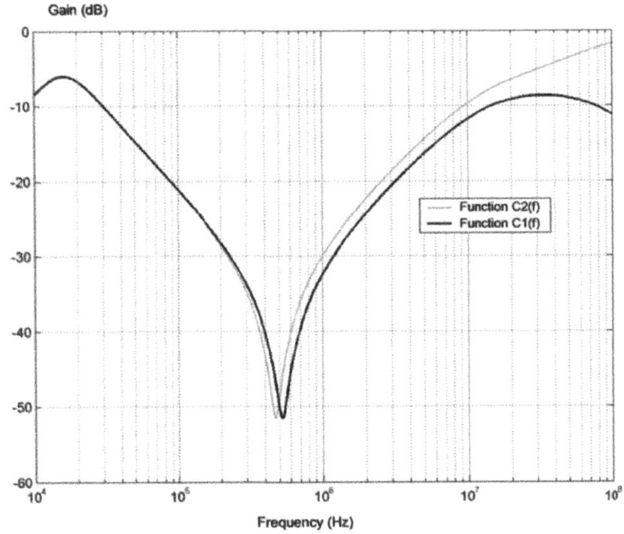

Figure 2.48. *Coupling functions $C_1(f)$ and $C_2(f)$*

2.7. Answers with comments

1) The expression of the impedance Z_L is given by:

$$Z_L = \frac{1}{jC_p\omega + \frac{1}{jL\omega} + \frac{1}{r_p}} = \frac{jL\omega}{1 + j\omega\frac{L}{r_p} + (j\omega)^2 LC_p}$$

As a Laplace operator s:

$$Z_L(s) = \frac{sL}{1 + s\frac{L}{r_p} + s^2 LC_p}$$

At low frequency for $F \ll \frac{1}{2\pi\sqrt{LC_p}}$, the impedance Z_L is only assimilable to the term *sL*.

At high frequency for $F \gg \frac{1}{2\pi\sqrt{LC_p}}$, the impedance is assimilable to the term $1/sC_p$.

The resonance frequency from which the change in behavior appears is:

$$F0 = \frac{1}{2\pi\sqrt{LC_p}} = 522.7 kHz$$

2) The equivalent ECM diagram is given in Figure 2.49. We have replaced the switching cell with its equivalent model consisting of linearized sources as discussed in section 2.2. The rectifier no longer appears since at any instant two diodes are available directly connecting the input of the converter to the LISN. We have added the parasitic elements of the passive components. We must note that the impedance Z_{RSIL} is very large and "insulates" the network of the converter relative to the HF quantities of the conducted perturbations. Thus, it appears that the main source is the voltage V_k, which contributes to the circulation of common- and differential-mode currents.

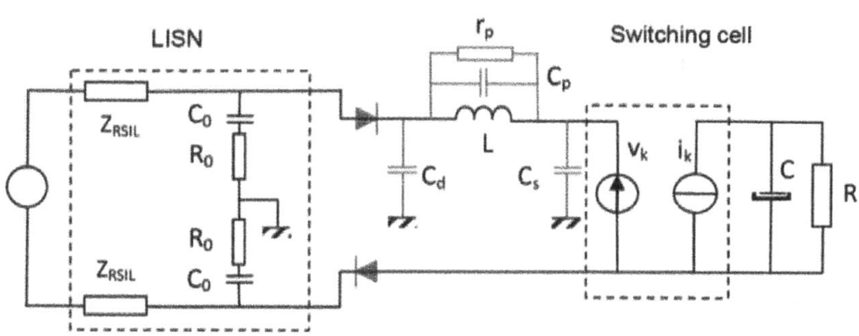

Figure 2.49. *Equivalent EMC model of the PFC*

About the inductance

The input inductance of the converter is an important element that contributes to the propagation of HF parasitic currents outside the converter, mainly due to the inevitable capacitive effects created in the windings and their proximity to the more or less conductive magnetic core. We can, therefore, represent these capacitive effects through the sectional diagram of the inductance, shown below. We must, nevertheless, keep in mind the idea that these phenomena are spread out.

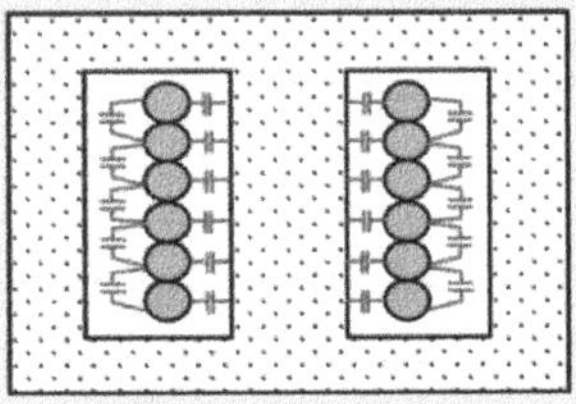

Sectional view of an inductance

The following curve shows the evolution of the impedance of an inductance as a function of the frequency, compared to the theoretical evolution.

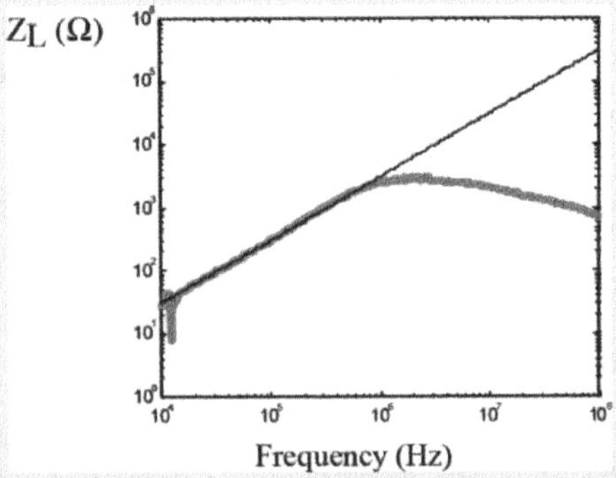

Variations in the impedance of a ferrite inductance as a function of the frequency

We note that from approximately 1 MHz onward, the curve changes slope, which is evidence of a capacitive effect. Finally, the losses in the materials (resistive loss in the conductors, losses through eddy currents in the materials, hysteretic magnetic losses) can be generalized for the whole model as a resistance, the value of which depends on the frequency F, the induction in the magnetic circuit, the properties of the magnetic material, the arrangement of the conductors in the magnetic circuit and their diameter. All of these complicated effects lead to the definition of an equivalent simplified electrical diagram for describing the behavior of the component at an operating point in the form of a parallel RLC circuit; other forms would also be possible.

3) The diagram can be simplified as shown below. We have not taken the current source i_k into account, as it does not contribute (in this simplified model) to the conducted perturbations undergone by the LISN. Moreover, the impedances Z_{LISN} are supposedly very large for HF parasitic currents, which do not traverse them here; they were, therefore, also removed from the simplified diagram.

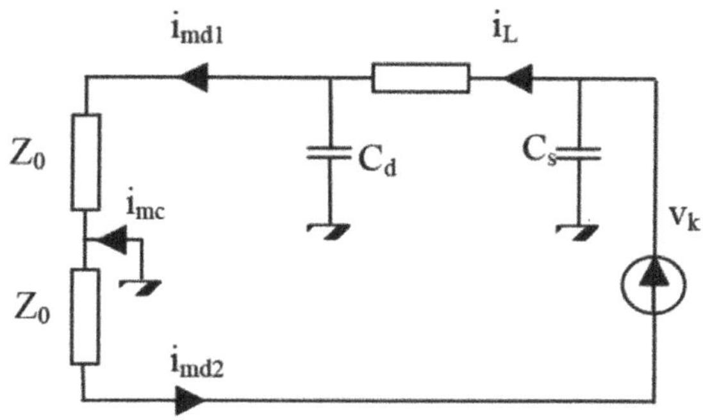

Figure 2.50. *Simplified differential mode model of the converter*

We let: $Z_0 = R_0 + \dfrac{1}{sC_0}$

4) The main source generating these HF perturbing currents in the LISN being v_k, we can, therefore, present it in the frequency domain. It is a supposedly ideal rectangular voltage, of amplitude Vs = 400 V, frequency 85 kHz and with a supposedly constant duty cycle equal to 0.5. We can give an analytical expression of the harmonic terms in Fourier series of this function:

$$v_n = 2V_s \cdot \frac{\sin(\alpha n\pi)}{n\pi}$$

where n is the harmonic order.

The shape of this function, obtained for the orders n such that the sine term is equal to 1, is therefore a decreasing function over 1/F (or −20 dB/dec). The asymptotic representation on a dBµV scale is given below.

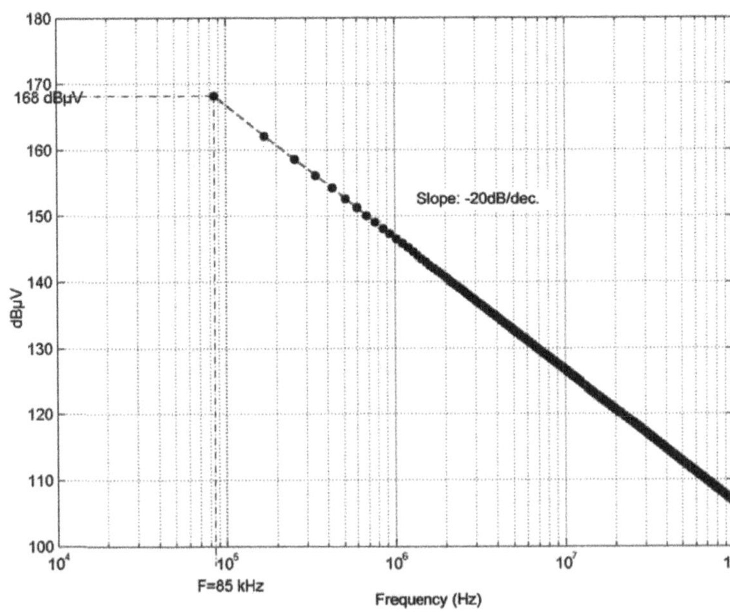

Figure 2.51. *Spectrum of the harmonics of a rectangular signal*

Observations on the spectrum of a non-ideal square wave

The general expression for the harmonics of a square signal with a rising time τ, duty cycle α and frequency F_0 is given by:

$$v_n = 2V_s\alpha \cdot \frac{\sin(\pi \cdot nF_o \cdot t_o)}{\pi \cdot nF_o \cdot t_o} \cdot \frac{\sin(\pi \cdot nF_o \cdot \tau)}{\pi \cdot nF_o \cdot \tau}$$

The first sine term is that of the previous expression, and determines the decrease at –20 dB/dec. The second can cause a second decrease at –40 dB/dec starting from the frequency 1/πτ. This second term acts in part as a filtering function, which "softens" the HF terms of the ideal square signal. We can, therefore, note that the larger τ is, the more the HF terms are softened.

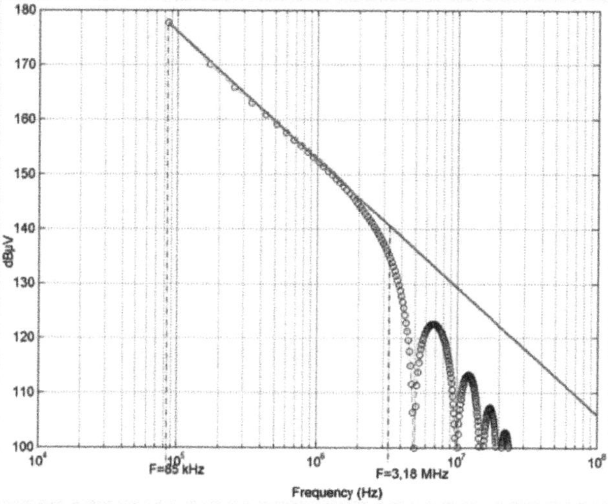

This result can be applied generally, and we show that a square signal with non-zero rise and fall times, the transition laws of which do not matter, can be expressed in the following way:

$$v_n = 2V_s\alpha \cdot \frac{\sin(\pi \cdot nF_o \cdot t_o)}{\pi \cdot nF_o \cdot t_o} \cdot \psi(F)$$

The function ψ(F) is the "smoothing" function of the transition; it determines the HF evolution of the harmonics of the real square signal. The higher its order of derivability, the faster the frequency harmonics are softened.

5) From the simplified diagram, it is possible to determine the different currents that traverse the LISN in common and differential modes; these are the perturbation currents created by this conversion structure.

We obtain the following formulas:

$$i_{md2} = \frac{V_k \cdot sC_0\left(\dfrac{1}{sC_s} + Z_L + \dfrac{1+sR_0C_0}{s(C_d+C_0)\left(1+s\dfrac{R_0C_0C_d}{C_0+C_d}\right)}\right)}{(1+sR_0C_0)\left(\dfrac{1}{sC_s} + Z_L + \dfrac{1+sR_0C_0}{s(C_d+C_0)\left(1+s\dfrac{R_0C_0C_d}{C_0+C_d}\right)}\right) + \dfrac{C_0}{C_s}\left(Z_L + \dfrac{1+sR_0C_0}{s(C_d+C_0)\left(1+s\dfrac{R_0C_0C_d}{C_0+C_d}\right)}\right)}$$

$$i_L = i_{md2} \cdot \frac{\dfrac{1}{sC_s}}{\dfrac{1}{sC_s} + \left(Z_L + \dfrac{1+sR_0C_0}{s(C_d+C_0)\left(1+s\dfrac{R_0C_0C_d}{C_0+C_d}\right)}\right)}$$

$$i_{md1} = i_L \cdot \frac{\dfrac{1}{sC_d}}{\dfrac{1}{sC_d} + \dfrac{1+sR_0C_0}{sC_0}} = i_{md2} \cdot \frac{\dfrac{1}{sC_s}}{\dfrac{1}{sC_s} + \left(Z_L + \dfrac{1+sR_0C_0}{s(C_d+C_0)\left(1+s\dfrac{R_0C_0C_d}{C_0+C_d}\right)}\right)} \cdot \frac{\dfrac{1}{sC_d}}{\dfrac{1}{sC_d} + \dfrac{1+sR_0C_0}{sC_0}}$$

$$i_{md1} = i_{md2} \cdot \frac{\dfrac{1}{sC_s}}{\dfrac{1}{sC_s} + \left(Z_L + \dfrac{1+sR_0C_0}{s(C_d+C_0)\left(1+s\dfrac{R_0C_0C_d}{C_0+C_d}\right)}\right)} \cdot \frac{\dfrac{C_0}{C_0+C_d}}{1+sR_0\dfrac{C_dC_0}{C_d+C_0}}$$

With:

$$V_{rsil1} = R_0 i_{md1}$$
$$V_{rsil2} = R_0 i_{md2}$$

$V_k = V_s \cdot f_m(t)$, we therefore identify the coupling functions $C_1(s)$ and $C_2(s)$ as:

$$C_1(s) = \frac{V_{rsil1}}{V_k}$$

$$C_2(s) = \frac{V_{rsil2}}{V_k}$$

Thus:

$$C_2(s) = \frac{sR_0C_0\left(\frac{1}{sC_s} + Z_L + \frac{1+sR_0C_0}{s(C_d+C_0)\left(1+s\frac{R_0C_OC_d}{C_0+C_d}\right)}\right)}{(1+sR_0C_0)\left(\frac{1}{sC_s} + Z_L + \frac{1+sR_0C_0}{s(C_d+C_0)\left(1+s\frac{R_0C_OC_d}{C_0+C_d}\right)}\right) + \frac{C_0}{C_s}\left(Z_L + \frac{1+sR_0C_0}{s(C_d+C_0)\left(1+s\frac{R_0C_OC_d}{C_0+C_d}\right)}\right)}$$

$$C_1(s) = \frac{sR_0C_0\left(\frac{1}{sC_s} + Z_L + \frac{1+sR_0C_0}{s(C_d+C_0)\left(1+s\frac{R_0C_OC_d}{C_0+C_d}\right)}\right)}{(1+sR_0C_0)\left(\frac{1}{sC_s} + Z_L + \frac{1+sR_0C_0}{s(C_d+C_0)\left(1+s\frac{R_0C_OC_d}{C_0+C_d}\right)}\right) + \frac{C_0}{C_s}\left(Z_L + \frac{1+sR_0C_0}{s(C_d+C_0)\left(1+s\frac{R_0C_OC_d}{C_0+C_d}\right)}\right)}$$

$$\cdot \frac{\frac{1}{pC_s}}{\frac{1}{sC_s} + \left(Z_L + \frac{1+sR_0C_0}{s(C_d+C_0)\left(1+s\frac{R_0C_OC_d}{C_0+C_d}\right)}\right)} \cdot \frac{\frac{C_0}{C_0+C_d}}{1+sR_0\frac{C_dC_0}{C_d+C_0}}$$

We deduce i_{mc}:

$$i_{mc} = i_{md2} - i_{md1} = \frac{V_k \cdot sC_0 \left(\frac{1}{sC_s} + Z_L + \frac{1+sR_0C_0}{s(C_d+C_0)\left(1+s\frac{R_0C_0C_d}{C_0+C_d}\right)} \right)}{(1+sR_0C_0)\left(\frac{1}{sC_s} + Z_L + \frac{1+sR_0C_0}{s(C_d+C_0)\left(1+s\frac{R_0C_0C_d}{C_0+C_d}\right)} \right) + \frac{C_0}{C_s}\left(Z_L + \frac{1+sR_0C_0}{s(C_d+C_0)\left(1+s\frac{R_0C_0C_d}{C_0+C_d}\right)} \right)}$$

$$\cdot \left(1 - \frac{\frac{1}{sC_s}}{\frac{1}{sC_s} + \left(Z_L + \frac{1+sR_0C_0}{s(C_d+C_0)\left(1+s\frac{R_0C_0C_d}{C_0+C_d}\right)} \right)} \cdot \frac{\frac{C_0}{C_0+C_d}}{1+sR_0\frac{C_dC_0}{C_d+C_0}} \right)$$

6) These functions are difficult to use analytically, so they have been digitally traced below with the given values, as presented in figure 2.52. We note that these functions have a minimum at a frequency close to 500 kHz, which corresponds to the parallel resonance of the inductance L (F_L = 522 kHz). At this frequency, the impedance of the inductance becomes very large, thus reducing the transmission of HF currents to the LISN.

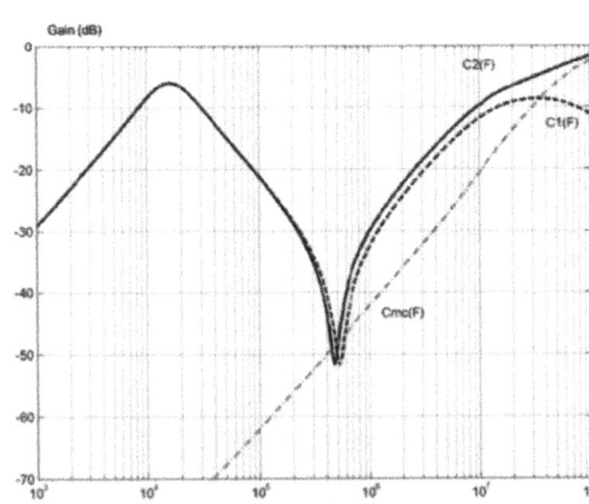

Figure 2.52. *Plotting of the different coupling functions of the converter*

We have also included the common-mode coupling transfer function $C_{mc}(F)$ defined as: $C_{mc} = R_0.i_{mc} / V_k$.

We can add that this function does not become particularly significant and only begins to dominate toward the higher frequencies. This effect is essentially dependent on the capacitance C_s and second on the capacitances C_d and C_p, with their weakness explaining this phenomenon.

7) At HF, the function $C_2(F)$ continues to increase since the impedance of the parasitic capacitance C_s becomes increasingly smaller and therefore allows the direct transmission of the current i_{md2} into the LISN, which gives us $i_{md2} = i_{mc}$. However, $C_1(F)$ decreases at HF since with the impedance of the capacitance C_d decreasing, the current i_{md1} in the LISN decreases; this parasitic capacitance, therefore, contributes to a kind of filtering. At low frequency (<10 kHz), the functions are identical and increasing, and the dominant effect is caused by the impedances L and Z_0 of the LISN (everything functions as though C_s and C_d were negligible). Thus, the functions increase between 1 and 10 kHz due to the capacitors in series C_0, and the maximum is reached for the series resonance between L and $C_0/2$ (16 kHz).

8) The resulting spectrum from the parasitic voltage V_{rsil1} measured by the LISN was plotted in Figure 2.53 (dark gray curve). We have also included the spectrum of the voltage V_k ($\alpha = 0.5$, $\tau = 200$ ns) as well as the coupling functions and the graph of the class A EN55011 standard.

We note that the spectrum of V_{rsil2} is softened in compliance with the coupling function $C_1(F)$, with a substantial minimum at the resonance frequency of the inductor L. In the frequency range from 600 kHz to 4 MHz, the spectrum is at a level that is almost constant: this results in the decrease over 1/F of the spectrum of V_k and the increase over F of the function $C_1(F)$ in this range. Beyond that, the decrease over $1/F^2$ of the spectrum of V_k and that of $C_1(F)$ in the 10–100 MHz range determines the decrease in the spectrum of V_{rsil1}.

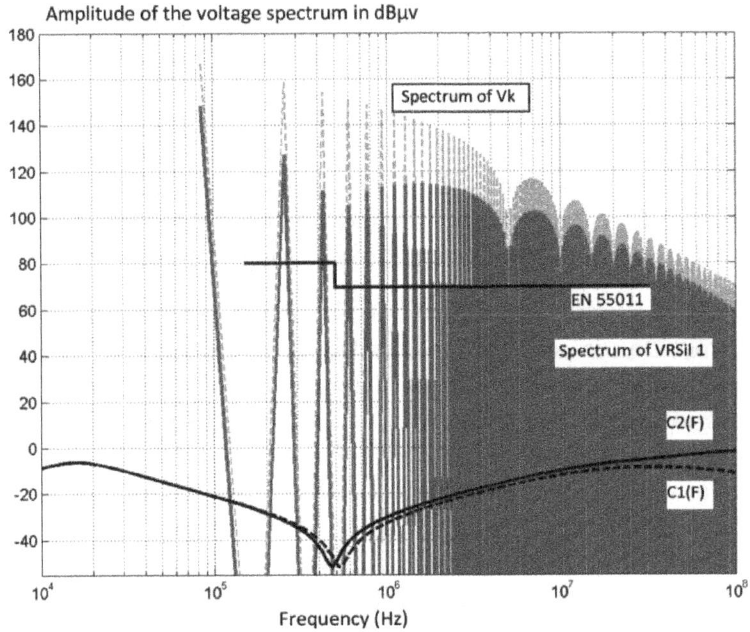

Figure 2.53. *Calculated spectra*

The observation of the measured spectrum lets us find these tendencies; we have superimposed the measured spectrum and the spectrum calculated from the model as shown in Figure 2.54.

The differences are explained by the following:

– The modulation function of the model was considered constant although in reality it varies between 1 and 0 over time, leading to the rise of even harmonics.

– The effects of the second-order parasites were not taken into account (connection inductances, output circuit, etc.) which explains the HF evolution of the spectrum (resonance peaks).

Nevertheless, we can see that such a simple model allows us to acceptably approximate the phenomena taking place.

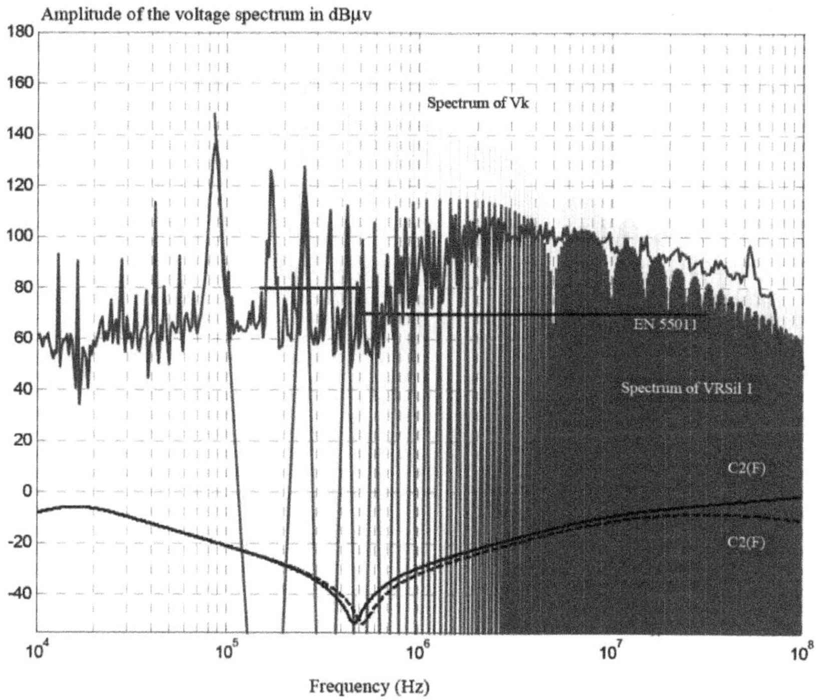

Figure 2.54. *Comparison with measured spectrum*

2.8. Bibliography

[COS 92] COSTA F., Contribution à l'étude des perturbations conduites dans les convertisseurs statiques haute fréquence, PhD Thesis, University Paris 11, 1992.

[COS 93] COSTA F., FOREST F., PUZO A., ROJAT G., "Étude des perturbations conduites et rayonnées dans une cellule de commutation", *Journal de Physique*, vol. 3, no. 12, pp. 2221–2248, 1993.

[CYR 01] CYRILLE G., Contribution au développement d'outils logiciels en vue de la conception des convertisseurs statiques intégrant la compatibilité électromagnétique, PhD Thesis, University Paris 6, 2001.

[PER 05] PERRET R., *Mise en oeuvre des composants électroniques de puissance*, Hermes-Lavoisier, Traité EGEM, p. 320, 2005.

[SCH 94] SCHANEN J.L., Intégration de la compatibilité électromagnétique dans la conception de convertisseurs en électronique de puissance, PhD thesis, University of Grenoble, 1994.

[SCH 00] SCHUSTER W.F.C., LEONHARDT G., "Electromagnetic simulation of bonding wires and comparison with wide band measurements", *IEEE Transactions on Advanced Packaging*, vol. 23, no. 1, pp, 69–79, 2000.

Chapter 3

EMC of Complex Electrical Energy Conversion Systems: Electromagnetic Actuators

3.1. How to define a complex system?

The first question that must be asked is: "what is a complex system?". The answer is not trivial and depends on the goals for the study of this system. In concrete terms, a complex power system joins different types of components which are all a part of the vast field of energy conversion. Thus, the association of static power converters, sensors, filters, controllers, actuators, such as electrical machines and of course mechanical or electrical loads, constitutes a complex set through the self-operation of each of these components. Each component requires and/or uses a set of voltages and currents of different frequencies and amplitudes. The interconnection, necessary for the global operation of the system, is at the origin of couplings which are often difficult to avoid, or at least to manage.

In order to illustrate the electromagnetic compatibility (EMC) aspects of this type of system, in this chapter, we propose focusing on a converter/electric AC machine association. This choice was influenced by the widespread use of these electromechanical trains in the industrial sector as well as in services and the household. The implementations are numerous, ranging from conveyor belts to axial flow units. The conversion chain dedicated to electric traction is a particularly interesting real-life example. The electrical, thermal, mechanical and EMC constraints by which this system for electric vehicles must abide define a rich and complex area of study.

The aim of this chapter is to show how such associations can have strong conducted emission power by analyzing the major sources and the most noteworthy means of propagation. Similar to the near-totality of electrical equipment aimed at European markets, variable-speed drives must conform to conducted and radiated EMC standards. Even if their widespread industrial use occasionally lets them avoid these constraints, their increasing use in services requires conforming to, or at least taking into account, the prescriptive aspects.

Variable-speed drive systems

Systems incorporating electrical systems represent more than 2/3 of industrial electrical energy consumption. Thus, 90 TWh of 120 TWh is consumed annually by motorized systems [ADE 03].

The industrial processes are numerous and the examples shown below are only a few among the many sectors that use variable-speed drives. Despite the different drawbacks, all the sectors have a common rule: managing their processes, notably using variable-speed drives. Paradoxically, these systems do not necessarily use electronic variable-speed drives. For example, only 3% of electrical machines with a power less than 2.2 kW are sold with a drive.

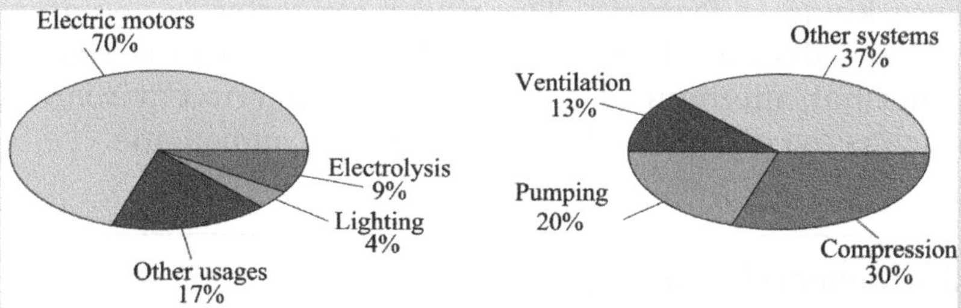

a) *Distribution of the consumption of electricity in the industrial sector in France*

b) *Distribution of the consumption of motorized systems*

3.2. Qualitative study

3.2.1. *Description of the conversion chain*

Before presenting a more detailed study, it is important to have a general view of the system and its constituent components. Drive systems, regardless of their use, have similar configurations. The electronic power converter (variable power drive) is an inverter with a voltage powered either by a rectified alternating source (i.e. a single-phase or a three-phase power grid connected or not to a controlled bridge rectifier) or by a direct current (DC) voltage (i.e. electrochemical battery). This inverter powers an electric machine with alternating current through a specific cable, the length of which depends directly on the implementation. Some of these, such as the industrial systems using submerged pumps, require that the variable-speed drive be removed from the machine and the electric connection has multiphase cables of a few meters to tens of meters. The electromagnetic impact of these very long cables is non-negligible and does not favor a decrease in global emissions conducted and radiated from the system. Embedded electric actuators specifically for civil and military aeronautics have these kinds of features. The power converter is voluntarily kept at a distance from the actuator as the latter is placed in a thermal, mechanical and even atmospheric environment, which is too disruptive for the electronic components.

More electric aircraft

The increased use of power electronics on aircraft is, on the one hand, linked to the decrease in consumption of non-propulsive power due to efficient and lighter electrical systems and, on the other hand, due to the reduction in maintenance costs without increasing the initial cost of these new systems. For example, on an A380, a portion of the hydraulic functions has been replaced by electric actuators. For the first time, the thrust reverser, which allows the thrust exerted by the reactor to be directed to the front, is actuated electrically by an electromechanical actuator (EMA). Moreover, the use of electrohydrostatic actuators (EHAs), which are local sources of hydraulic energy provided by an electric compressor, simplifies the workings of flight controls. The figure below illustrates the use of EHAs in the field of aeronautics.

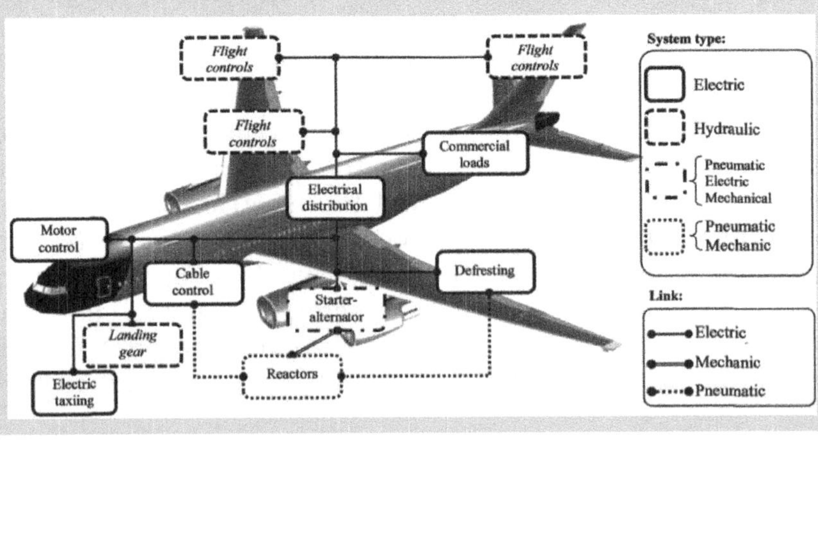

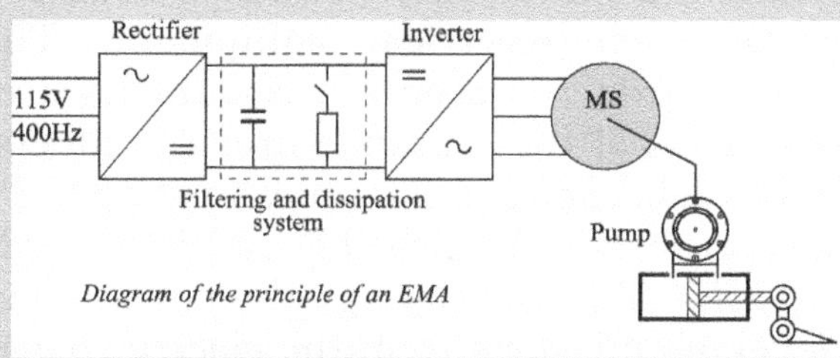

Diagram of the principle of an EMA

Photo of an EMA

The wheel motors dedicated to electric taxiing are another example of a now embedded electrical system. It uses an auxiliary power unit (APU) in order to function. This system lets aircraft move along the runways without having to use the main motors. Power electronics and its control give the pilot complete control over the speed, direction and braking of the aircraft on the ground.

The different parts of this conversion chain can be considered either as local sources of perturbations or as prioritized propagation tracks. If we were to consider a simplified and standardized study, the tested system would consist of an ideal source of DC voltage perfectly isolated from the earth, a line impedance stabilizer network (LISN) (see Chapter 1), an EMC filter (see Chapter 4), a three-phase inverter, the cooling of which is handled by a radiator, a power cable and a three-phase electric machine (Figure 3.1).

3.2.2. *Reminder of the standards*

As we have seen in the previous chapters, the analysis and understanding of EMC problems for converters or all other electronic device are implicitly guided by the different

European and international standards, nowadays required for the marketing of such electrical equipment. The conformity to EMC standards represents a certain level of quality and competitiveness of the manufacturer's products. This conformity, therefore, becomes a true sales incentive. To back up this analysis, we will reference the CENELEC EN 61800-3 standard [CEN 04], specifying the measuring conditions and the level of emissions from the electric drives and the variable-speed drive. The levels are given on a logarithmic scale in dBμV [3.1]; this scaling system will, therefore, be used from now on for all representations of spectrums.

$$dB\mu V(v) = 20 \cdot log_{10}\left(\frac{v}{10^{-6}}\right) \quad [3.1]$$

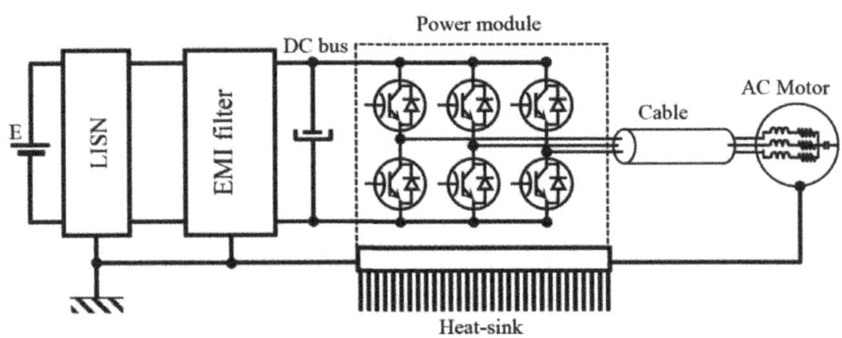

Figure 3.1. *Structure being studied*

The specifications of the standards relative to the perturbations conducted in the radio frequency band (150 kHz–30 MHz) are split into two categories [CEN 04]. The first category, known as class A, defines the level of emissions for appliances used in the industrial sector. The second category, and certainly not the least, is reserved for the household and hospital sectors: this is known as class B. As we can see (Figure 3.2), the latter has a much more restrictive profile than that of the former, the levels of which are recalled in Table 3.1.

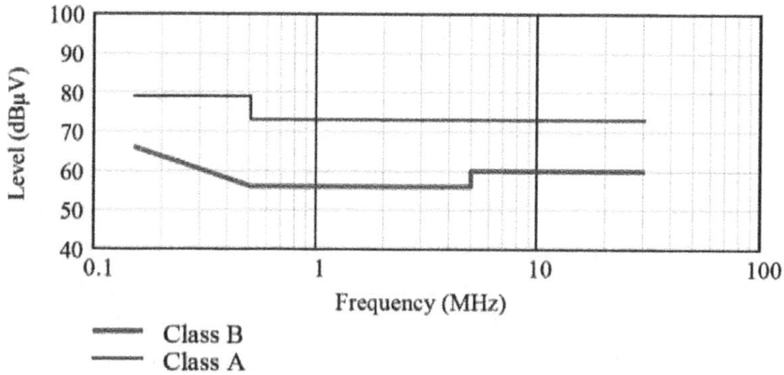

Figure 3.2. *Levels of conducted emissions for EN61800-3*

Frequencies (MHz)	Class A (dBµV)	Class B (dBµV)
0.15–0.5	79	66–56
0.5–5	73	56
5–30	73	60

Table 3.1. *Limits of the profiles for classes A and B*

Generally, the designers of electrical equipment look to conform to the criteria of class B, with this allowing them to cover a wider market, and by alleviating the filtering methods, to cater only to class A.

3.2.3. *Propagation methods*

3.2.3.1. *Common and differential modes*

Like all power structures, this converter–machine association presents two conventional propagation methods for conducted emissions: common and differential modes. The connection scheme between the earth and the different parts is critical as it defines the route of the common-mode currents of this set. The currents take the paths with the least impedance which prioritize capacitive connections at high frequencies. Strictly speaking, it would not be necessary to distinguish between these two modes since the standardized tests are carried out with no differences at the LISN level (Figure 3.3). In practice, this breakdown is often, even systematically, carried out in order to define which passive or active filtering strategy is preferable to implement. This analysis can be carried out upstream from the chain (grid side) as well as downstream (machine side). Nonetheless, if the electrical load of the system is not considered to be particularly vulnerable to perturbations or does not require any specific study, only the standardized analysis of the power supply is retained. In other words, the LISN is placed between the power supply and the converter.

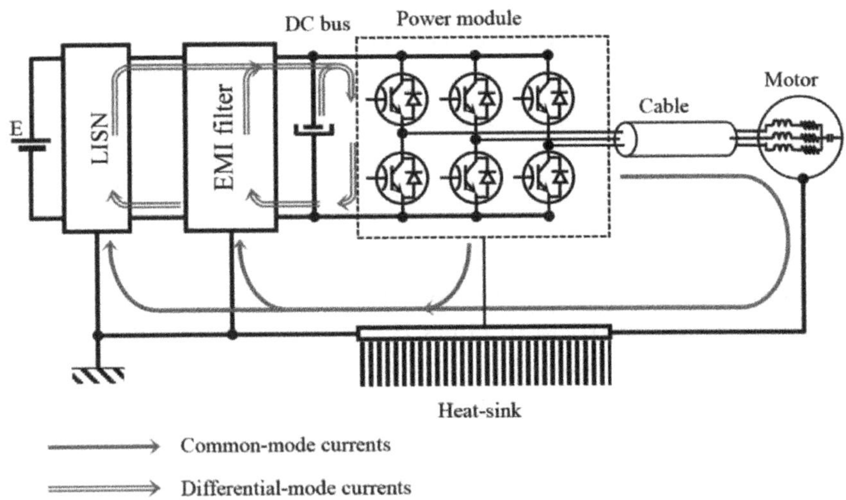

Figure 3.3. *Flow of the perturbation currents*

3.2.3.2. *The power module and its neighboring elements*

The power modules connect a set of elementary chips in the same package in order to directly address the desired operating function. They offer a simple assembly as opposed to discrete components, the number of which can quickly become prohibitive for structures with several switching cells. They are used in low- and medium-power industrial variable-speed drives, and typically incorporate a set of six isolated gate bipolar transistor (IGBTs) and their anti-parallel diodes [SEM 13]. An input rectifier can equally be used in this model. Certain modules, known as intelligent power modules (IPMs), even incorporate local control drives and protection devices. The silicon chips are soldered onto copper tracks which are placed on the same fine ceramic substrate (≈ 600 μm). The backside is also made of copper to improve the thermal conductivity of the module which facilitates exterior exchanges (Figure 3.4).

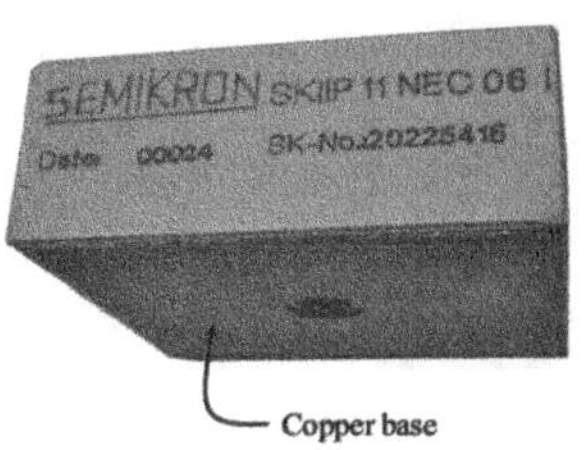

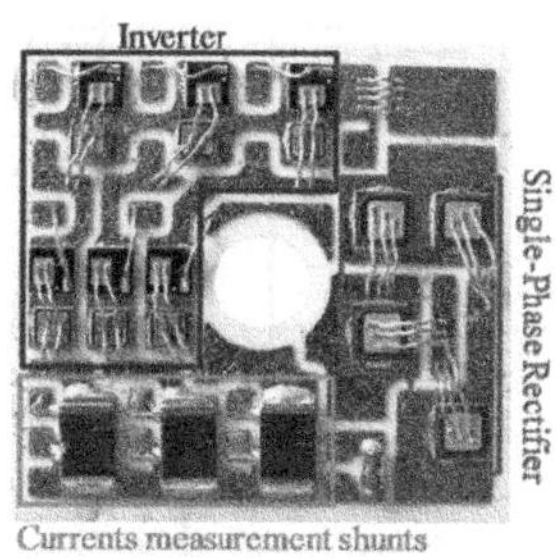

Figure 3.4. *IGBT power module (Semikron 600 V –17 A). For a color version of this figure, see www.iste.co.uk / costa / electronics.zip*

Because of this, the silicon chips and the different copper tracks are found within direct proximity to an earth potential through an intermediate earthed radiator for security reasons. This has the direct effect of capacitive coupling between the tracks placed on the potential of the collectors and the rear surface of the module. We saw in the previous chapter that these couplings modeled by parasitic capacitances are the source of common-mode currents in power structures. The value of these parasitic elements can be relatively weak, which renders their measurements very approximate (<15 pF on the model tested in Figure 3.4). Orders of magnitude can then be established on the basis of traditional analytical relations, knowing the shape of the copper tracks in the module. The surface of the chip varies depending on the current rating which naturally induces a modification of the copper track on which it travels. The capacitive couplings are, therefore, entirely dependent on the placement of chips in the framework (Figure 3.5).

The compactness of these modules would be a bonus for managing undesirable electromagnetic phenomena if filtering elements (i.e. decoupling capacitors) were also incorporated, which sadly is not the case yet. These essential filtering cells are instead moved to the outside of the module, thus limiting their efficiency.

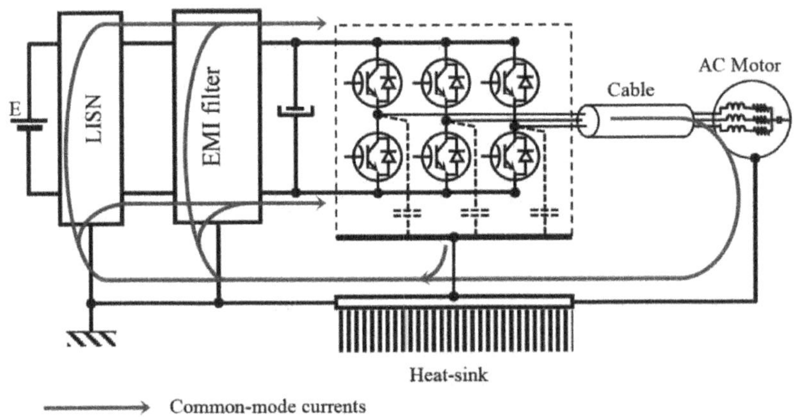

Figure 3.5. *Capacitive couplings of the model and flow of the common-mode currents*

3.3. Modeling in frequency domain

3.3.1. *Linearization of the switching cell*

The aim of EMC modeling is to approach the spectrum of perturbation signals on the frequency range relative to the conducted emissions as defined by the standards. Therefore, it is interesting, although not ideal, to work directly in the frequency domain. Nevertheless, the direct frequency-based methods sometimes require additional experience than those which use time-based simulations. For a given structure, they require knowledge of the major mechanisms of perturbation generation on the one hand and knowledge of the set of critical propagation channels on the other hand, which can quickly prove to be complicated. There is, however, a simplified process which is not strictly in line with the theory but which has the advantage of helping to understand the phenomena, conduct sensitivity analysis and finally take on the problem of design. This relatively simple process rests upon the breakdown of different perturbation sources and propagation channels. By taking the descriptions given in the previous chapters, the sources can be broken down into differential-mode (current sources) and common-mode (sources of voltage). We recall that this disassociation of the effects of current and voltage variations allows for a simple and quick analysis of phenomena. This method is completely based on the "source-path-victim" process; however, it requires a good knowledge of the power structure.

The conducted perturbations, whether common-mode or differential-mode, in the voltage inverter, like the static converters presented earlier, mainly originate from the switching transients of power switches generating strong *dv/dt* and *di/dt*. The simplest method used in this chapter, therefore, consists of replacing the switching cells by current and voltage sources which represent the image of switched electrical quantities caused by the state change of the switches (Figure 3.6) [REV 11]. Relation [3.2] lets us prove that an inverter leg generates a voltage between the midpoint of the latter and the earth. This voltage, denoted by v_M, is directly dependent on the voltage at the terminals of the switches. It represents one of the common-mode voltages which will generate parasitic currents throughout the impedances defined by the cable and machine.

We have $\begin{cases} v_M = v_H - v_{k1} \\ v_M = v_L + v_{k2} \end{cases}$

$$\text{With} \begin{cases} v_{DC} = v_H - v_L \\ v_{k1} = v_{DC} - v_{k2} \end{cases}$$

$$v_M = \frac{v_H + v_L}{2} + \frac{2 \cdot v_{k2} - v_{DC}}{2} \qquad [3.2]$$

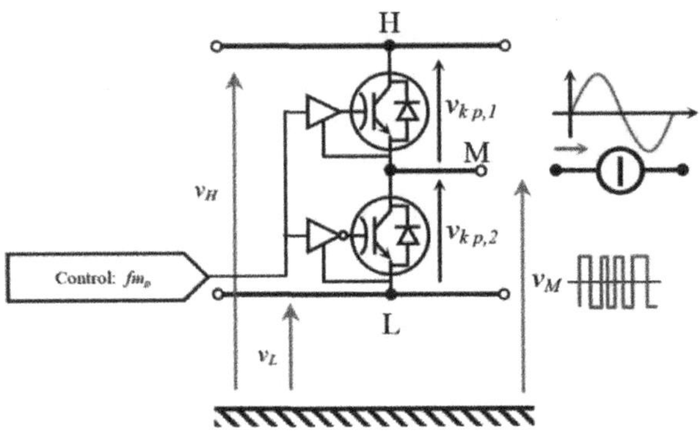

Figure 3.6. *Inverter leg*

The control strategy of the power switches of the leg p of the inverter can be emulated by a binary function f_{m_p}(modulation function), which, based on its value, will set the state of the switches [3.3]. The voltage at the ends of the switch $k_{p,2}$ is directly expressed through the product of f_{m_p} and v_{DC}, the DC bus voltage. The voltage at the midpoint of the inverter branch v_M is thus evidently dependent on the modulation strategy [3.5].

$$f_{m_p} = \begin{cases} 1, \{\mathrm{k_{p,1}closed, k_{p,2}\ open}\} \\ 0, \{\mathrm{k_{p,1}open, k_{p,2}closed}\} \end{cases} \qquad [3.3]$$

$$v_{k\,p,2} = f_{m_p} \cdot v_{DC} \qquad [3.4]$$

$$v_M = \frac{v_H + v_L}{2} + \frac{(2 \cdot f_{m_p} - 1)}{2} \cdot v_{DC} \qquad [3.5]$$

Following a similar process, the analysis of the currents of the DC bus upstream of the cell, let us define a differential-mode current (Figure 3.7). For this, the mode-decoupling hypothesis is necessary, which supposes that only differential-mode currents exist. By applying the usual definition of a differential-mode current [3.6], the current i_{MD} is given by relation [3.7].

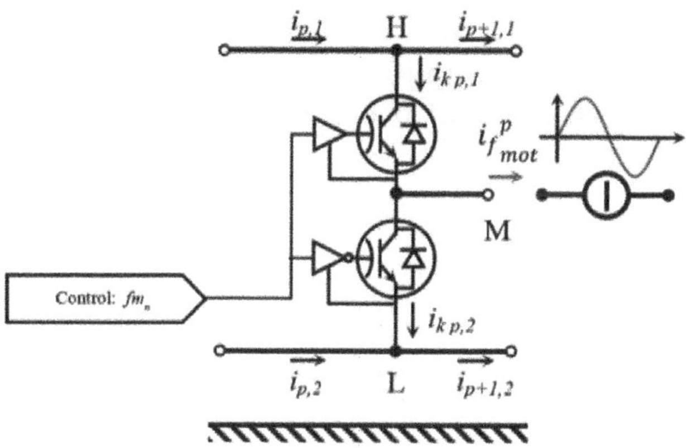

Figure 3.7. *Switching cell seen in differential mode*

$$i_{MD_p} = \frac{i_{p,1} - i_{p,2}}{2} \quad [3.6]$$

$$\text{With} \begin{cases} i_{p,1} = i_{k\,p,1} + i_{p+1,1} \\ i_{p,2} = -i_{k\,p,2} + i_{p+1,2} \end{cases}$$

$$i_{MD_p} = \frac{i_{p+1,1} - i_{p+1,2}}{2} + \frac{i_{k\,p,1} + i_{k\,p,2}}{2} \quad [3.7]$$

The currents in the switches are naturally complementary and can be expressed in terms of $i_{f\,mot}^{p}$ which represents the current of the phase p of the machine. The expression of the total differential-mode current i_{MD} is, therefore, expressed in terms of the switched currents of each inverter leg $i_{k\,p,1}$ and the currents $i_{f\,mot}^{p}$ [3.8].

$$i_{MD} = \sum_{p=1}^{3} \left(\frac{2 \cdot i_{k\,p,1} - i_{f\,mot}^{\,p}}{2} \right) \qquad [3.8]$$

In order to simplify this expression, and to finally obtain a clear relation, we can note that in the absence of common-mode currents (hypothesis of the separation of perturbation sources) the sum of the three phases of the $i_{f\,mot}^{\,p}$ terms is naturally zero for a non-earthed machine. The resulting relation of this simplification [3.9] shows that the differential-mode current is expressed in terms of the sum of switched currents $i_{k\,p,1}$.

$$i_{MD} = \sum_{p=1}^{3} i_{k\,p,1} \qquad [3.9]$$

The current in the switches is directly linked to the load current $i_{f\,mot}^{\,p}$ via the modulation function f_{m_p} [3.10].

$$i_{k\,p,1} = f_{m_p} \cdot i_{f\,mot}^{\,p} \qquad [3.10]$$

Thus, the differential-mode currents created by each inverter leg, caused by the load and control strategy, can be modeled by a single current generator i_g [3.11].

$$i_g = \sum_{p=1}^{3} f_{m_p} \cdot i_{f\,mot}^{\,p} \qquad [3.11]$$

This technique allows us to linearize the operation of the inverter legs by replacing the switches with equivalent generators, the waveforms of which are known. The three inverter legs are thus modeled by an equivalent current generator representing the switched current traveling through the DC bus and by three equivalent voltage generators representing the output voltages of the inverter (Figure 3.8).

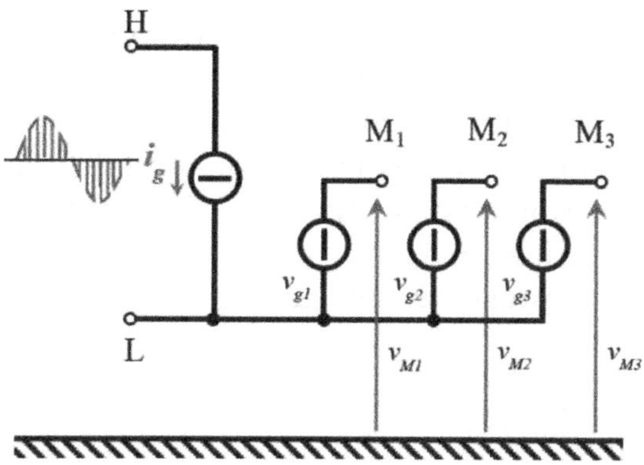

Figure 3.8. *Equivalent linear diagram: equivalent generators*

3.3.2. *Modeling of the perturbation sources*

3.3.2.1. *Definition of the frequencies*

As opposed to DC-DC converters, variable-speed drives, and more generally voltage inverters, use driving laws which make switching moments evolve over time in order to modulate the power signals. The difficulty in correctly representing the spectrum of a pulse width-modulated (PWM) signal resides in the definition of the frequencies that make it up and which implicitly correspond to the duration over which the signal is studied. To simplify the proposed approach, the switching frequency is defined as a whole multiple of the frequency of stator currents (Fs). The obtained PWM is, therefore, said to be "synchronous" and the true period of the signal does, indeed, correspond to that of the motor that we wish to enforce. In the opposite case, this frequency can no longer be used as the reference for the calculation of the sources spectrum. As the Fourier transforms, which we will present later on, remain rigorous, it is necessary to determine a calculation step corresponding to a sub-multiple of both the switching frequency and the desired motor frequency.

$$F_{dec} = M \cdot \delta f \qquad M \in \mathbb{N} \qquad F_S = N \cdot \delta f \qquad N \in \mathbb{N}$$

In order to reduce the number of frequencies involved as much as possible, and therefore calculation time, it is preferable to use the largest common frequency between F_{dec} and F_s which then corresponds to the actual frequency of the signal in question. Thus, the calculation frequency δf is defined by relation [3.12] in which $gcd(x,y)$ represents the function that represents the greatest common divider of x and y.

$$\delta f = gcd(F_{dec}, F_S) \qquad [3.12]$$

If this condition is not met, the frequency analysis would not be carried out over a whole number of periods and the Fourier transform of the involved signals would present additional harmonics. This can be clearly shown in a simple example. Let a signal $s(t)$ consist of two frequencies f_1 and f_2 of 50 Hz and 1,030 Hz, respectively, and the amplitudes of which are arbitrarily set to 1 [3.13].

$$s(t) = \sin(2\pi \cdot f_1 \cdot t) + \sin(2\pi \cdot f_2 \cdot t) \qquad [3.13]$$

First, the large gap between the two frequencies could justify a study of the frequency of this signal at 50 Hz; however, we can see in Figure 3.9 that of the two frequencies usually present, only that of 50 Hz is correctly presented. With a chosen frequency resolution of 50 Hz, the second frequency cannot be correctly represented. In addition, rays appear around f_2 at frequencies for which they do not usually exist. The actual frequency step δf of this signal is equal to 10 Hz. A frequency analysis carried out with this 10 Hz frequency will naturally help to obtain the expected spectrum (Figure 3.9).

The problems encountered in simulations linked to transient states lead to strictly identical phenomena. As long as the signal has not reached a steady-state periodicity, the spectral analysis is not valid. This reasoning can also be extended to measurements by trying to capture signals over a correctly defined duration. In reality, the steady state is never truly maintained since the control system “constantly” modifies the operating point of the converter. These variations are fortunately relatively weak and measurements carried out over sufficiently long periods of time (several periods) let us indirectly create an average of the values, thus minimizing errors from disparities in measurements [POP 99].

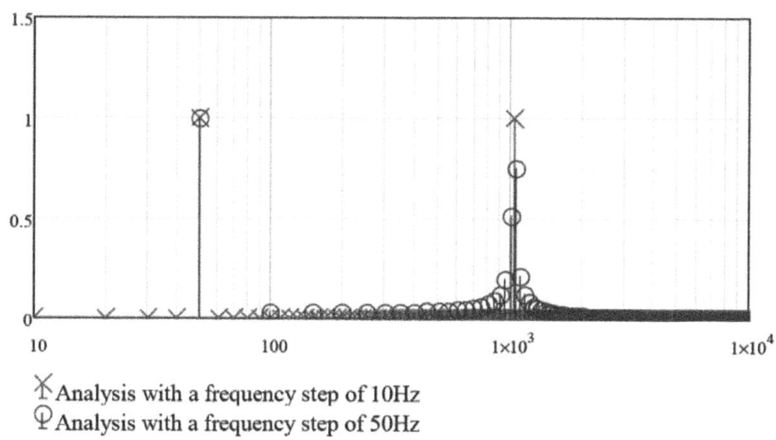

Figure 3.9. *Influence of the analysis frequency over the spectrum*

3.3.2.2. *Control laws: pulse-width modulation*

The control techniques of inverters are numerous. However, it is possible to group them into families and identify, for example, methods for the direct calculation of conduction durations for switches, or for the placement of vectors representing output voltages due to a combination of switch states.

To summarize, two large families of PWM are distinguishable:

– The first family groups the large set of strategies which allow the evolution of pulse widths from instantaneous values. Be they local (control of a cell) or global (vector management of several cells), the instantaneous PWMs are generally used in applications where the switching frequency (F_{dec}) is large compared to that of the fundamental (F_S).

– The second family encompasses the techniques through which waveforms are calculated over a fundamental period, memorized and then used where necessary during an operation. These modulations are called calculated PWMs.

The numerous modulation techniques do not allow an exhaustive study of them, which is why we will limit ourselves to certain cases that ultimately are sufficiently representative. Indeed, we will expose the notable differences between simple "natural" modulations and "space-vector" modulations used for the variable-speed drive.

They have the advantages of being sufficiently explicit and intuitive to analytically define the spectral representation of the voltages at the terminals of switches. It is for these different reasons that we will not limit our study to only one family.

3.3.2.3. *Natural PWM*

We will start with the simplest scenario of natural modulations, which consist of comparing two signals of differing frequencies in order to define a pulsed signal of fixed frequency and varying duty cycles. They allow for decentralized management of the control. In other words, leg (i.e. switching cell) of the inverter is controlled by its own distinctive modulating signal.

The signal tasked with transporting the "information" used by the load in the case of an inverter is called a "carrier". It is typically created from triangular or sawtooth signals. The information in question (modulating signal) originates from a signal, the load of which we wish to impose a temporal form; this is called the "modulating" signal. Although they are generally considered to be less efficient and limited in terms of the control of systems, contrary to vectorial strategies, they are still widely used in fixed-frequency applications such as uninterruptible power supplies (UPS).

The two waveforms used as carriers, presented below (Figure 3.10), are traditional ones, owing to their increasing "triangle" and "sawtooth" shapes. To simplify their description, these functions have a unitary amplitude and they are centered at zero. They allow us to distinguish two natural classes of PWM. The first, defined by the triangle shape, causes moments of switching-on and switching-off, which vary from one switching period to another. If we once again consider a very large switching frequency compared to that required by the load (i.e. the machine), then the modulating function can be regarded constant which indicates control orders that are symmetrical relative to the peaks of the triangles. In the case of a sawtooth carrier, whether increasing or decreasing, one of the two control orders is inevitably defined by a multiple of the switching period. This is a case of natural PWM centered to the left or the right.

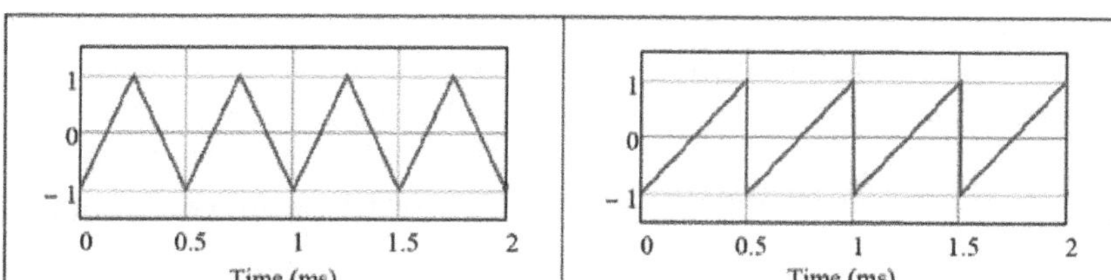

a) *Example of a triangular carrier at 2kHz* | b) *Example of a sawtooth carrier at 2kHz*

Figure 3.10. *Examples of carriers*

Their influence on the spectrum of a pulse-width modulated signal and unitary amplitude is, therefore, significant. Figure 3.11 presents a small part of the spectrum around the switching frequency set at 4 kHz. We can observe, in the traditional scenario of a sinusoidal modulating signal, a large increase in multiple odd low-frequency rays, normally limited by the effects of symmetry.

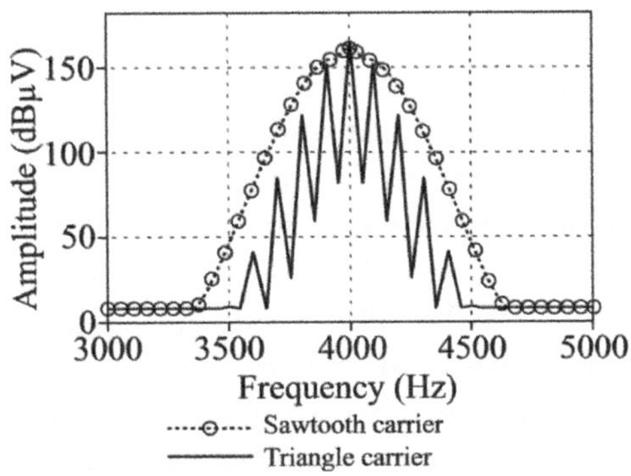

Figure 3.11. *Influence of the carrier on the spectrum of a sinusoidal modulation at 50 Hz*

For any given modulation law, the maximum rays, which define the outer envelope of the spectrum, are roughly the same regardless of the type of carriers used. This observation suggests that the impact of sawtooth carriers or of any other asymmetrical shape on conducted perturbations is minimal. This conclusion is, however, excessive as the density of the spectrum on low-frequency bands, distinctive of a PWM, contributes through standardized measuring protocols (spectral analyzer) to the increase in the level of the measured spectrum. Indeed, the input filter of the spectral analyzer, required during certification testing, modifies the envelope of the spectrum by cumulating the rays, the spacing of which is less than the selected bandwidth.

3.3.2.4. *Space-vector PWM*

Vectorial PWMs appeared with the rise in machine-control techniques using alternating currents and they are linked to the use of different base-changing operators (Park, Concordia, etc.). They allow for a global control of the inverter legs contrary to natural PWMs. Control through space-vector PWM consists of controlling the inverter through a vector defined in the Concordia space, the inverse transformation of which is given by relation [3.14]. The coordinates of this control vector are functions of elementary vectors each associated with a switch state.

$$\begin{pmatrix} v_\alpha \\ v_\beta \end{pmatrix} = \sqrt{\frac{2}{3}} \cdot \begin{pmatrix} 1 & -{}^{1}/_{2} & -{}^{1}/_{2} \\ 0 & -{}^{\sqrt{3}}/_{2} & -{}^{\sqrt{3}}/_{2} \end{pmatrix} \cdot \begin{pmatrix} v_{1n} \\ v_{2n} \\ v_{3n} \end{pmatrix} \qquad [3.14]$$

With v_{in}: the voltages at the terminals of the machine.

With three legs, each having only two states, the number of possible configurations for this type of inverter increases to 8 (2^3). Each of these configurations has a vector representing the image of the voltages of the load in the Concordia space. The plane in the Concordia space is then divided into sectors, the limits of which are determined by the vectors V_i. These are the transformations of the eight configurations of the modulation functions f_{mi}, which help define the simple voltages of the machine in the space $\mathcal{R}(\alpha,\beta)$ (Figure 3.12).

$$\begin{pmatrix} v_{1n} \\ v_{2n} \\ v_{3n} \end{pmatrix} = \frac{V_{DC}}{3} \cdot \begin{pmatrix} 2 & -1 & -1 \\ -1 & 2 & -1 \\ -1 & -1 & 2 \end{pmatrix} \cdot \begin{pmatrix} f_{m1} \\ f_{m2} \\ f_{m3} \end{pmatrix} \qquad [3.15]$$

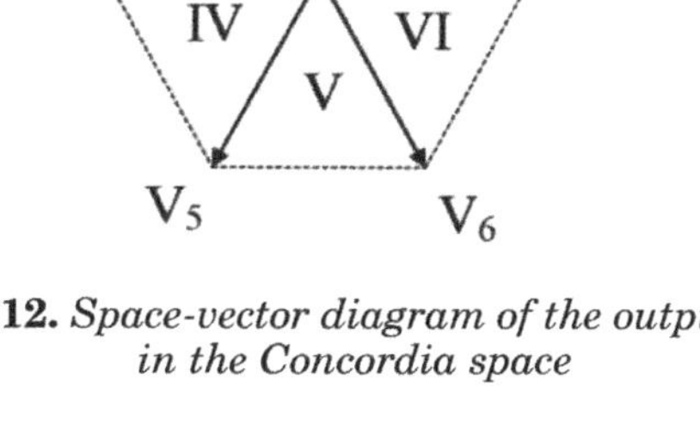

Figure 3.12. *Space-vector diagram of the output voltages in the Concordia space*

The aim is to create an average vector compared to the switching representing the desired output voltages by making a linear combination of two elementary vectors which define the limits of the quadrant in which it is located. Thus, we define the conduction duration that the switches must take over the switching period being used. To illustrate this process and understand how these switching moments are determined, we have chosen a traditional configuration in which the switching patterns are centered on the half-period of the switching. Figure 3.13 shows how the vector $V_{\alpha 1\ \beta 1}$ is constructed in the quadrant due to its projections on the vectors V_1 and V_2. The succession of states, represented in Figure 3.13, is an example of a strategy which consists of centering the control patterns of the switches on the half-period T_{dec}. The series of vectors with odd and even states allows us to reduce the number of commutations to 2 per leg on a period T_{dec}.

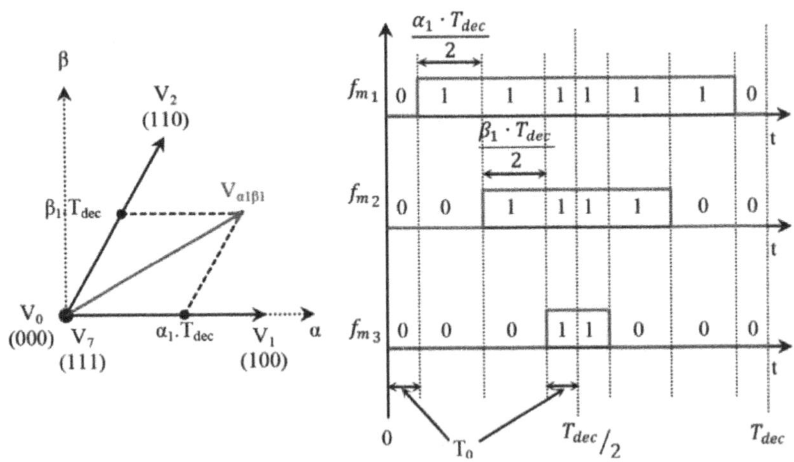

Figure 3.13. *Example of the construction of the sample vector in the first quadrant*

Because of this description, we can see that the conduction time of the switches can be expressed in terms of the duration of each state. These different times are determined by introducing the notion of average value of voltages on the switching period. Some mostly proven formulas [GUI 98] give the conduction time for each switch directly, on the three branches of the inverter, regardless of the current PWM period.

3.3.2.5. *Definition of the voltage sources*

The signals defined by a pulse-width modulation have a spectral makeup that is much more complex than the traditional waveforms such as square waves and trapezium waves. However, over a switching period, these quantities are similar since the temporal shape of the voltage at the terminals of a switch always corresponds to the switching of a continuous voltage. This is naturally the sum of a high number of elementary signals, with one of their parameters evolving over time which considerably modifies the spectral density. In the case of usual modulations, it is the evolution of the pulse-width of the control voltage from one switching period to another which begets low-frequency components according to the control strategy. The output voltage from the conversion, therefore, contains the frequency components of the control law. In order to represent what we have defined as the output voltage of the inverter legs (Figure 3.14), and from this the sources of perturbations of the inverter, we must properly define the evolution of the square waveforms over time.

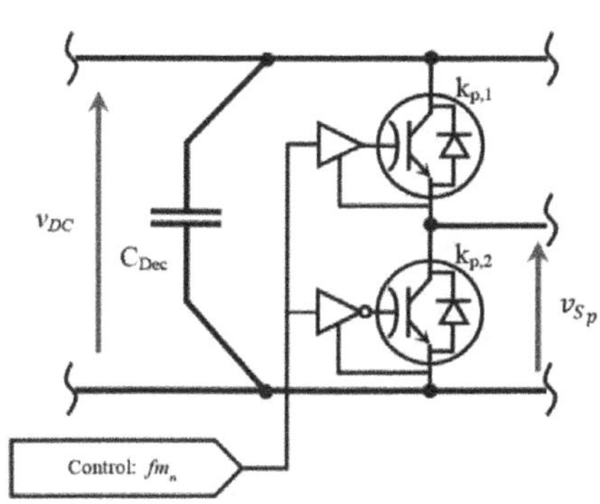

Figure 3.14. *Output voltage of an inverter branch*

The switching moments governed by the control strategies, which we will come back to, change from one switching period to another. Starting from the principle that there always exists an even number of commutations on the studied period, these essential dates can be classified following the type of commutation with which they are associated (opening, closing) (Figure 3.15). Thus, for the ith studied period, we will denote by $t_{on_{p,i}}$ the instant corresponding to the switching-on of the transistor and by $t_{off_{p,i}}$ that relative to the switching-off of the transistor.

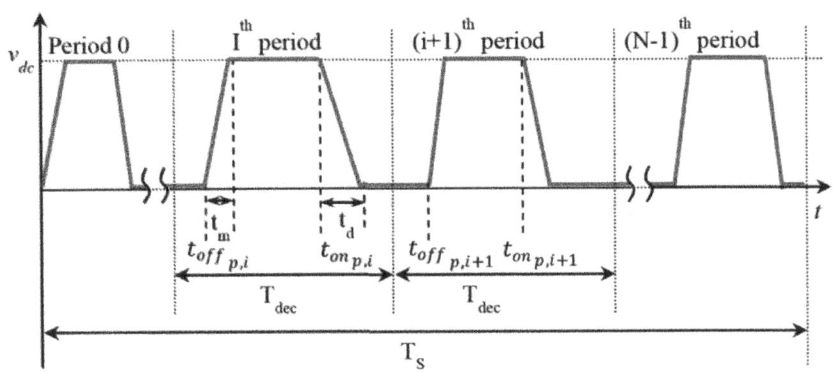

Figure 3.15. *Modulated trapezoidal signal*

The harmonic decomposition over the whole studied period (T_S) is provided by expression [3.16], where N represents the whole number of switch-mode periods (T_{dec}) contained in the studied window of time (T_S). The variable n defines the multiple of the frequency of the analysis F_S ($T_S{}^{-1}$) (i.e. harmonic). The choice of this studied time window, over which the frequency representation is determined above, is not arbitrary and depends heavily on the major operating frequencies such as the switching frequency or the frequency of the stator currents which we try to set indirectly. They correspond to the true period of the signal.

$$V_{g_{p,n}} = \begin{cases} 2 \cdot V_{dc} \cdot F_S \cdot \sum_{i=0}^{N-1} \left(\frac{t_d - t_m}{2} + t_{off_{i,p}} - t_{on_{i,p}} \right), & n = 0 \\ \frac{V_{dc}}{\pi \cdot j \cdot n} \cdot \sum_{i=0}^{N-1} \left(sinc\left(\frac{t_m \cdot n}{T_S} \right) \cdot e^{-\pi \cdot j \cdot \frac{(2 \cdot t_{off_{p,i}} + t_m)}{T_S} \cdot n} - sinc\left(\frac{t_d \cdot n}{T_S} \right) \cdot e^{-\pi \cdot j \cdot \frac{(2 \cdot t_{on_{p,i}} + t_d)}{T_S} \cdot n} \right), & n \neq 0 \end{cases}$$

[3.16]

Before we consider the above relation, we can make the qualitative observation that the sum does not influence the frequency dynamics of the spectrum's envelope. Furthermore, the cardinal sine values corresponding to the rise and fall times, respectively, t_m and t_d have no influence at the lowest frequencies since their parameters remain very weak. The spectrum, therefore, has a decrease which is inversely proportional to the frequency, at 20 dBµV per decade. For higher frequencies, the denominator of the cardinal sine values intervenes by increasing the negative slope of the spectrum. Thus, like the majority of theoretical spectrums of voltages at the terminals of power switches, the amplitude of the rays decreases with the frequency. In other words, with this signal having been created from a superimposition of a square waveform, it is absolutely normal that its spectrum presents similar characteristics.

Figures 3.16 and 3.17 present the spectrum of the voltage at the terminals of a power transistor of an inverter operating with a natural sinusoidal PWM and a space-vector PWM, respectively. These two PWMs are adjusted so that the amplitude of the fundamental (50 Hz) of the calculated voltage is equal to 100 V.

3.3.2.6. *Current source*

The load current is set by the motor windings via the mechanical load of the shaft. In the case of voltage inverters, the output current of a leg is alternating. Furthermore, the motor connected to variable-speed drives is essentially a heavily inductive load for low frequencies which naturally filter currents from the inverter. The current $i_{cel\ p}$, defined as the current from a switching cell, corresponds to the output current switched by the same commutation function as V_s, the voltage at the terminals of the switch $k_{p,2}$ (Figure 3.18).

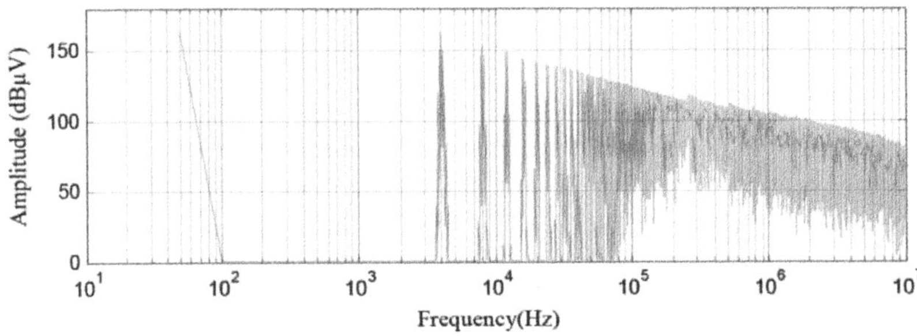

Figure 3.16. *Representation of the spectrum of a voltage source modulated by a natural sinusoidal PWM for $V_{dc} = 300$ V, $F_{dec} = 4$ kHz and $F_S = 50$ Hz*

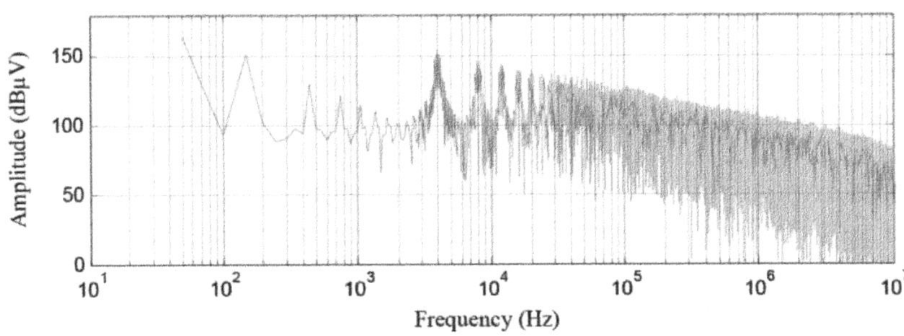

Figure 3.17. *Representation of the spectrum of a voltage source modulated by a space-vector PWM for $V_{dc} = 300$ V, $F_{dec} = 4$ kHz and $F_S = 50$ Hz*

By connecting the three legs of the inverter, the sum of the cell's currents allow us to only represent one current source which is connected in parallel at the input of the inverter. This source is then defined as a function of the output currents and the control law of the switches of each leg. Like the voltage sources, the current source must be defined in the frequency domain in order to preserve the advantages of a direct simulation. Although the definition of the voltage sources is relatively simple due to the sum of the modulated square waves, the current source seems more complicated since the modulated square waves have an amplitude which evolves over time.

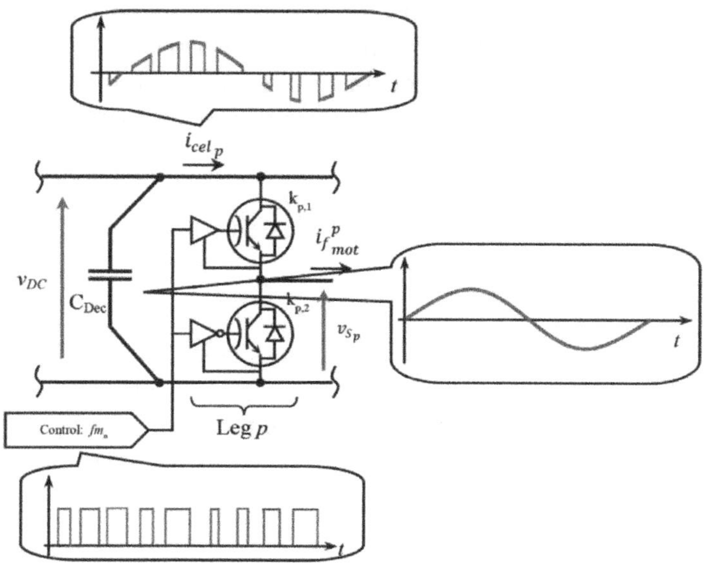

Figure 3.18. *Sign conventions for the output voltage and current of a switching cell*

To start, we need the Fourier transform of an elementary trapezoid with a unitary amplitude relative to the ith switching period. This transformation, deduced directly from [3.17], is also expressed in terms of the output phase number (p) of the inverter. Therefore, variables $t_{off_{i,p}}$ $t_{on_{i,p}}$ correspond, respectively, to the switching-off and switching-on moments in the period i and for the phase number p (p = {1 2 3}). Times t_d and t_m correspond to the fall and rise times of the currents, i.e. the switching durations for the switching-on and switching-off.

$$Tr_i^p(\upsilon) = \begin{cases} 2 \cdot F_S \cdot \left(\frac{t_d - t_m}{2} + t_{off\,i,p} - t_{on\,i,p}\right) & , \quad \upsilon = 0 \\ F_S \cdot \frac{sinc(t_m \cdot \upsilon) \cdot e^{-\pi \cdot j \cdot (2 \cdot t_{on\,i,p} + t_m) \cdot \upsilon} - sinc(t_d \cdot \upsilon) \cdot e^{-\pi \cdot j \cdot (2 \cdot t_{off\,i,p} + t_d) \cdot \upsilon}}{\pi \cdot j \cdot \upsilon} & , \quad \upsilon \neq 0 \end{cases}$$

$$\text{With} \quad sinc(x) = \frac{sin(x)}{x} \qquad [3.17]$$

Therefore, it is necessary to use the Fourier transform of the sinusoidal output currents of the inverter [3.18] and to make use of the properties of the Dirac distribution which considerably simplifies the calculations in the frequency domain.

$$I_{f\,mot}^{\;p}(\upsilon) = \frac{I}{2 \cdot j} \cdot \left(\delta(\upsilon - F_S) \cdot e^{-j \cdot \left((p-1) \cdot \frac{2\pi}{3} + \varphi\right)} - \delta(\upsilon + F_S) \cdot e^{j \cdot \left((p-1) \cdot \frac{2\pi}{3} + \varphi\right)}\right) \qquad [3.18]$$

The input current of a switching cell, such as we have defined it, corresponds to the product of the the modulation function of the switching cell with the output current. Therefore, it is a convolution which we must carry out in the frequency domain [3.19].

$$I_{cel\,i}^{\;p}(\upsilon) = I_{f\,mot}^{\;p}(\upsilon) * Tr_i^p(\upsilon) \qquad [3.19]$$

With the Dirac distribution being the neutral element of the convolution, the result is worked out quickly without any intermediary calculations. The harmonics (n) of the source of perfect perturbations (I_G) is also given by the following relation:

$$I_{G\,n} = \frac{I}{2 \cdot j} \sum_{p=1}^{3} \sum_{i=0}^{N-1} Tr_i^p\left(\frac{n-1}{T_S}\right) \cdot e^{-j \cdot \left((p-1) \cdot \frac{2\pi}{3} + \varphi\right)} - Tr_i^p\left(\frac{n+1}{T_S}\right) \cdot e^{j \cdot \left((p-1) \cdot \frac{2\pi}{3} + \varphi\right)} \qquad [3.20]$$

Outlining the exact spectrum is nevertheless extremely costly in terms of numeric calculations since it requires a double sum to be carried out for each frequency. However, due to modeling by equivalent perturbation sources, as long as the operating point of the inverter does not change, the calculation is only carried out once. The spectrum of the perturbation source which we have just defined does not take the deformations associated with the control law into account. Indeed, this technique supposes that the load perfectly filters the harmonics of the modulated square waves in order to preserve the fundamental. If this hypothesis is verified for a sinusoidal modulation, it will quickly be proved false for more complex laws. In the case of a natural PWM, the theoretical knowledge along with precision in the moments of switching allows the generation of a spectrum, the contents of which are almost null for low frequencies (Figure 3.19) as opposed to a space-vector PWM which, following the desired strategy, can bring out a richer spectrum (Figure 3.20).

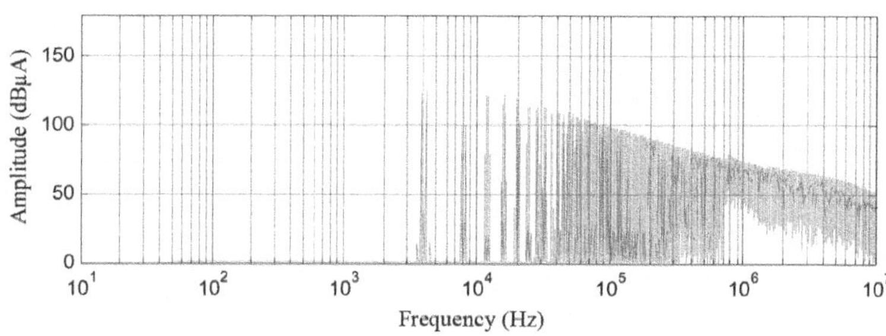

Figure 3.19. *Spectral representation of the current source for a current of 10 A and $\varphi = -30°$ and a natural PWM at 4 kHz*

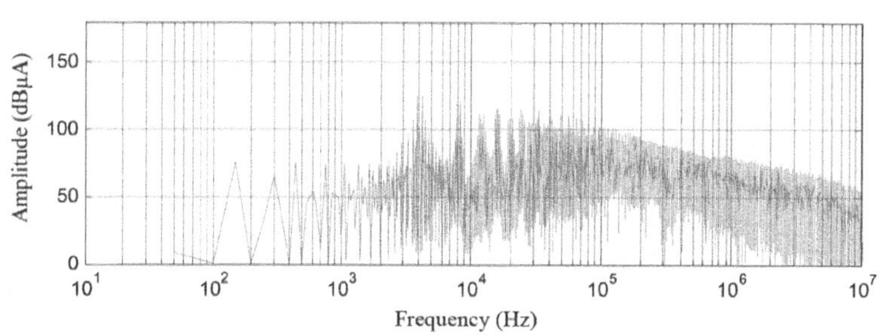

Figure 3.20. *Spectral representation of the current source for machine currents of 10 A and $\varphi= -30°$ with a space-vector PWM at 4 kHz*

The analytical formulas in use here are exact and the precision of the obtained results in the frequency domain is independent of the number of frequencies being handled. Nevertheless, the frequency-based approach is evidently not restrictive. Due to an appropriate number of frequencies ($N = 2^P$, $P \in \mathbb{N}$), the use of the inverse fast Fourier transform (FFT) is possible, which lets us obtain the major values in the time domain as well (Figure 3.21). Although the calculated spectrum is exact, however, it is not infinite and can cause small oscillations in the signal where there are discontinuities. This is of course the Gibbs phenomenon, which can be particularly difficult to avoid. The equivalent generator connecting the three switch-mode currents through the legs allows us to obtain a rectified current, an example of which is provided by the inverse FFT in Figure 3.22.

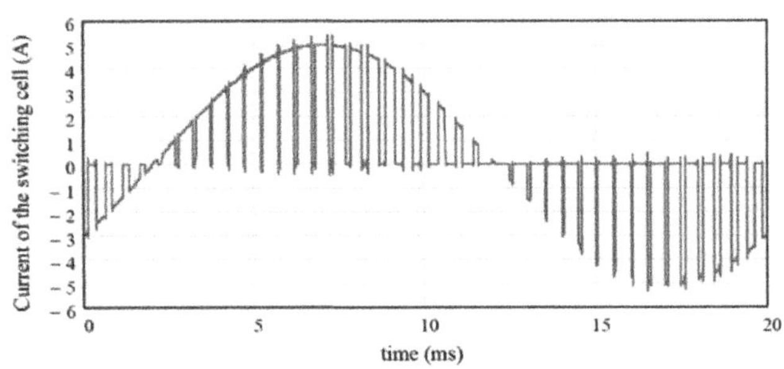

Figure 3.21. *Inverse FFT of the current of a leg for a switch-mode frequency of 2 kHz (machine displacement factor of 0.8)*

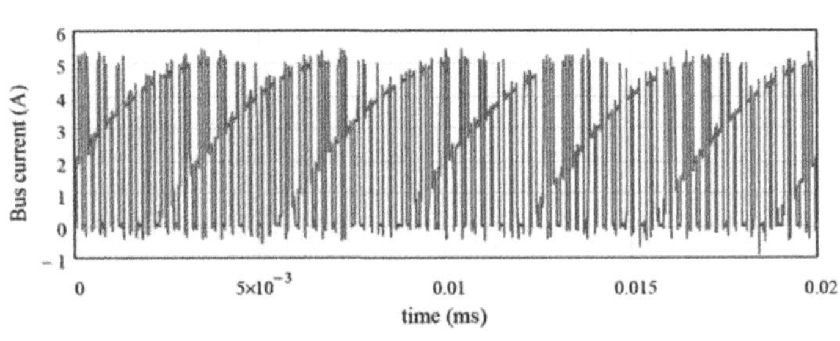

Figure 3.22. *Temporal reconstitution of the current rectified by the inverse FFT of the exact spectrum for a given frequency*

3.4. Frequency-based representation of an inverter

A three-phase PWM inverter, at its core, is a common-mode current generator. Despite a three-phase load made up of motor windings, which is supposedly perfectly balanced, there is always a residual neutral voltage which causes high common-mode currents via a set of parasitic capacitive couplings. These perturbations are contained due to propagation channels either purpose-built for this effect (EMC filter and/or LISN) or that are undesirable (power supply). All of the complexity in EMC modeling, therefore, resides in the definition of models which are both simple and consistent over the chosen frequency range. Therefore, they must guarantee the realism of simulations, while limiting risks in instability. The aim of this section is to outline a circuit equivalent to this type of structure capable of identifying the number of perturbations in an acceptable fashion.

3.4.1. *Equivalent common-mode source – simplified diagram*

The three legs of the inverter constitute the major sources of perturbations of this structure. Specifically, these sources are modeled by three harmonic voltage generators. The numerous works previously carried out in this area of research [HOE 01, CON 96] prove the predominance of common-mode currents and ignore the influence of the differential mode on the frequency range relating to the conducted perturbations. This observation allows us to outline the extremely simple equivalent frequency-based model of the structure. This model, presented in Figure 3.22, incorporates the three voltage generators as well as the propagation channels defined by the load (motor and cable) which we will revisit later. The LISN is also represented by the impedance Z_{rsil} (Figure 3.23). However, this circuit is based on the hypothetical symmetry of the inverter in terms of common-mode propagation channels and we suppose from this that the total common-mode current is equally spread in the two branches being measured of the LISN. These branches are, therefore, connected in parallel. The load and the set of earth couplings are represented by an equivalent phase impedance.

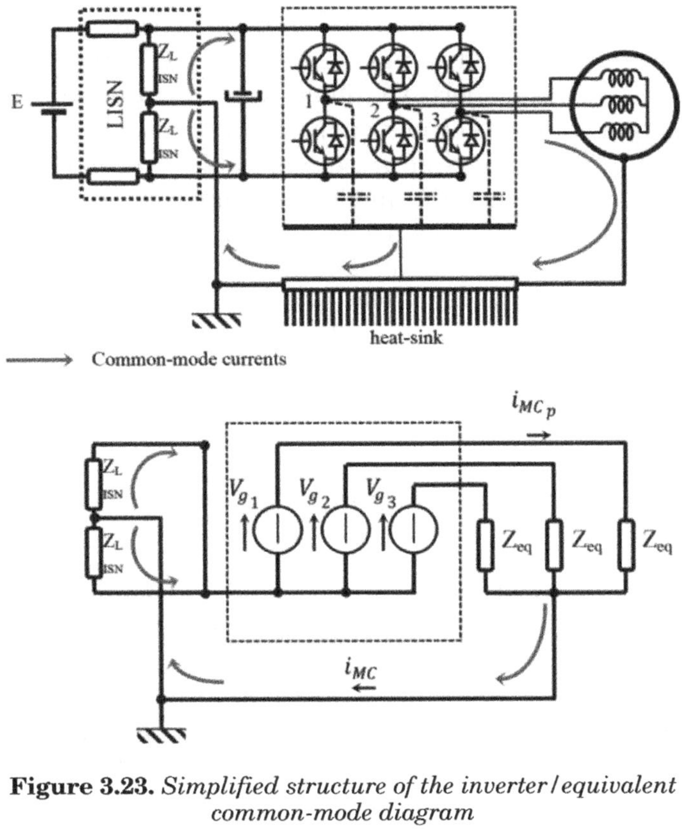

Figure 3.23. *Simplified structure of the inverter / equivalent common-mode diagram*

The sum of the three relations outlined in equation [3.21] allows us to express the total common-mode current in terms of the generators and the major propagation channels in this same mode. The resulting relation brings a level of additional simplicity by transposing the three-phase diagram [3.22] into a single-phase which, in retrospect, seems relatively obvious. Thus, like the impedances of the LISN, the elements of the load are found in parallel. The unique common-mode source V_{MC} is, therefore, equivalent to the "average" of the three initial generators.

It is obvious that the model used here is extremely simplified, and it allows for a quick and simple estimation of the common-mode currents of a structure which, originally, seemed a relatively complex task. It also shows the major common-mode source which simply represents the neutral voltage. This source, defined by the superimposition of the set of modulated square waves, is a multilevel voltage, the amplitude of which lies between –E/2 and E/2 with E being the DC bus voltage (Figure 3.24).

$$V_{g_p} = Z_{eq} \cdot i_{MC_p} + \frac{Z_{RSIL}}{2} \cdot i_{MC} \qquad \text{With p= 1,2,3} \qquad [3.21]$$

$$i_{MC} = \frac{1}{\frac{Z_{eq}}{3} + \frac{Z_{RSIL}}{2}} \cdot V_{MC} \qquad \text{with} \qquad V_{MC} = \frac{1}{3} \cdot \sum_{p=1}^{3} V_{g_p} \qquad [3.22]$$

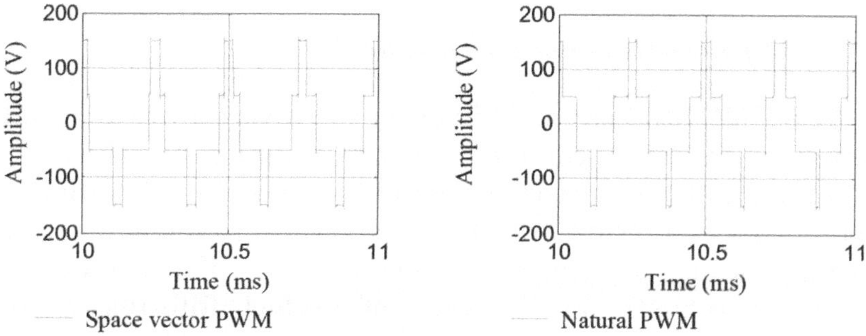

Figure 3.24. *Time-based representations of a section of the common-mode voltage*

In the case of a traditional space-vector PWM, presented in the previous paragraphs, the waveforms of the common-mode source resemble those of a sinusoidal natural PWM since the modulated square waves are centered on the switch-mode period. Nonetheless, the spectrum associated with this neutral voltage provides distinctly more harmonics than we could see in Figure 3.25. For this simulation, the modulation rate of the sinusoidal natural PWM is fixed so that the fundamental of the output voltage, over a phase of the inverter, has an identical root mean square to that provided by a space-vector PWM (e.g. 100 V).

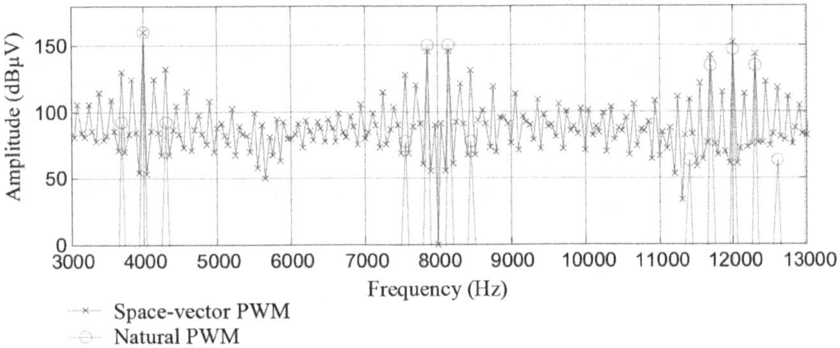

Figure 3.25. *Representation of a section of the spectrum of the common-mode voltage*

3.4.2. *Differential-mode influence*

The EMC studies carried out on the inverter are generally focused on the common mode, supposing that the differential-mode perturbations only have a small influence over the spectrum studied on the frequency range corresponding to standard benchmarks. This viewpoint, although realistic in the case of perfect filtering, is too restrictive and does not let us model the filtering elements in the best way. A study similar to the one conducted for the common mode can be adopted, this time by using the current source defined previously. In this scenario, the whole active part of the inverter is replaced by this current source representing the current traveling through the DC bus (Figure 3.26). Therefore, it is essential to include the equivalent capacitor of the DC bus necessary to its filtering without which the majority of harmonics would end up in the LISN. This element synthesizes one or more decoupling capacitors of the DC bus.

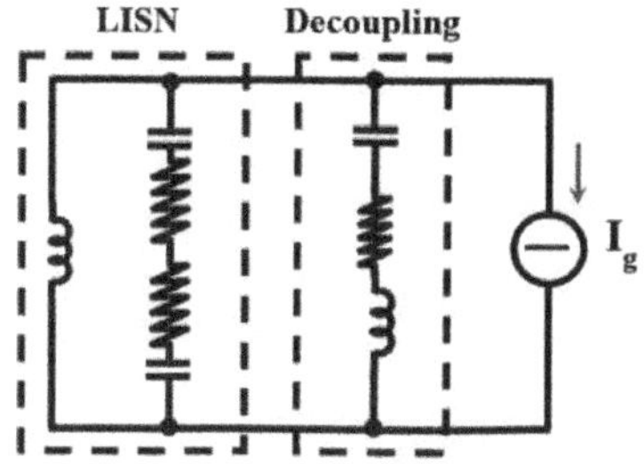

Figure 3.26. *Equivalent differential-mode diagram*

Power: DC bus PWM	1kW 300V sinusoidal (80%)	
Equivalent capcitor		C = 1mF R = 50mΩ L = 50nH

Table 3.2. *Parameters of the model*

In the absence of a differential-mode filter, it is mainly the decoupling capacitor of the DC bus which conditions the level of perturbations detected by the elements of the LISN. The harmonics generated by the power currents of the motor and the switchings are filtered by this high-value component. Unfortunately, its efficiency is considerably hampered by its intrinsic parasitic components. The value of the total capacity, designed to withstand strong and efficient currents, and the high levels of voltage of the DC bus require electrochemical technology for its components (Figure 3.27). For this simple and realistic example, due to the different values chosen, the level of the differential-mode spectrum is tangent to the profile of the class B standard, and although it is significantly smaller than the common-mode equivalent, it is critical compared to the level required (Figure 3.28).

Figure 3.27. *Capacitive decoupling of an industrial variable-speed drive*

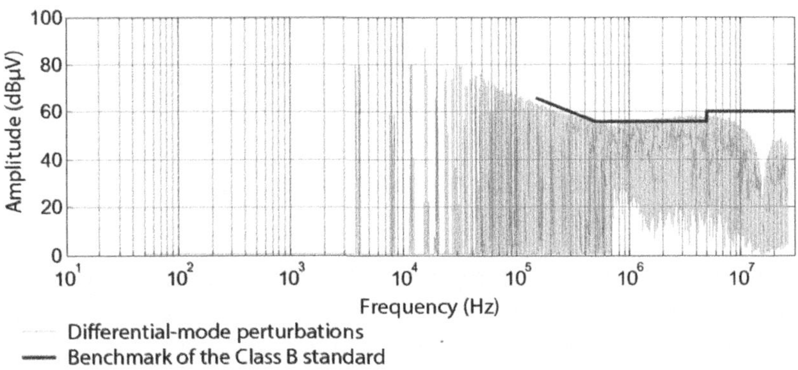

Figure 3.28. *Differential-mode spectrum*

3.4.3. *Proposed frequency-based diagram*

A three-phase inverter, from a topological point of view, is a symmetrical converter. In other words, the parasitic common-mode currents do not, in principle, have prioritized channels and flow throughout the structure to eventually return to the origin where they were created. This symmetry is usually located within the switching cells which, as opposed to those of the series chopper which we have already presented, have two identical switches with the same function and complementary controls. The frequency-based model is shown in Figure 3.29. This model incorporates the LISN, one level of filtering, the components of the DC bus, the perturbation sources and the load consisting of the machine and its power cable.

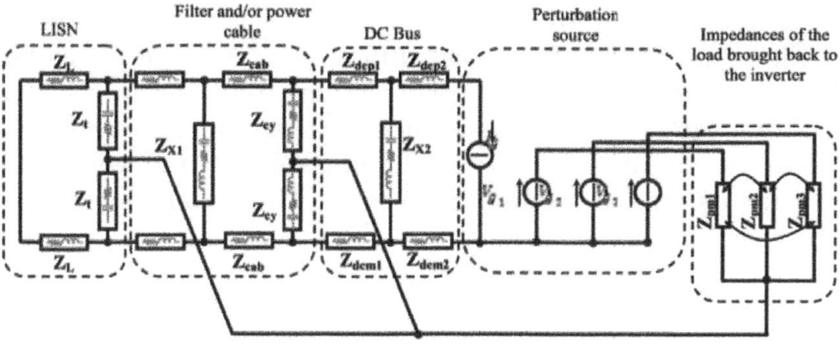

Figure 3.29. *Equivalent frequency diagram*

The power components, incorporated into the module, present capacitive links to the earth, which are connected in parallel with the load (machine + cable). All of these impedances are grouped together and define the Z_{pm} components of the equivalent diagram.

3.5. Modeling of the cables and motors

3.5.1. *Estimation of the primary parameters of the power cables*

In this configuration, the length of the cable is not a parameter to be ignored with regard to the frequencies being used. Indeed, the propagation phenomena appear for frequencies at which the corresponding wavelength becomes less than the length of the conductors. The propagation phenomena cause overvoltages at the motor and converter levels which can reduce their lifespan, or in the worst case they cause the destruction of the equipment [CAC 99]. The subject of this study is a shielded three-phase cable with an extra conductor for earth protection. The use of shielded cables becomes necessary in order to limit radiated effects. It also allows for the diversion of part of the common-mode current traveling to the motor.

Because of this, the shielding causes a significant increase in the capacitance between the different phases and the earth. The experimental measurements taken from the impedance analyzer HP4194A help us to identify the inductance and the linear capacitance of this shielded cable plaited with four conductors with a section of 4.5 mm^2. These components are presented in Table 3.3.

	5 kHz	30 MHz
L (nH/m)	328	197
C (pF/m)	295	262

Table 3.3. *Cable parameters measured through impedance analyzer HP4194A*

Although the conductors are made up of strands with a thin section (0.1 mm^2), the induced effects (skin and proximity effects) cause a variation in the measured values in terms of frequency.

For outlining an estimation of the electrical values RLCG and simulating the behavior of the frequency of a power cable of a given length, an execution of a geometric model of a multi-conductor shielded cable is required. The different primary parameters introduced earlier due to the measurement can also be calculated using different techniques (partial element equivalent circuit (PEEC) analytical expression of cylindrical conductors, etc.)

Shielded cables have been the subject of numerous studies [PAU 92] which have resulted in analytical expressions of the different parameters that we are looking for. Thus, for N conductors of radius R_C protected by a perfectly conductive shielding, relations [3.23] and [3.24], respectively, give self- and mutual inductances as a reference conductor. These relations are given in terms of the geometric parameters presented in Figure 3.30. These expressions obviously rest on the presumption that the space is homogeneous and can be characterized by unique efficient dielectric and magnetic constants.

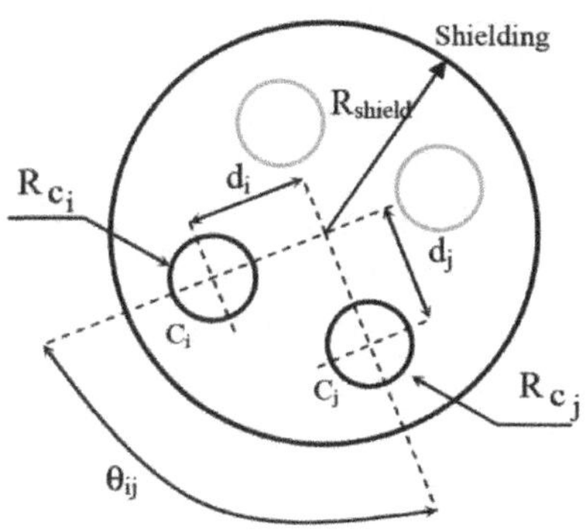

Geometric parameters of a cable ($4\times4mm^2$ 1000V)	
R_{shield}(mm)	5.5
R_C (mm)	1.13
d_i (mm)	2.7
θ_{ij} (degrees)	90

Figure 3.30. *Geometric outline of a power cable*

These relations are established by supposing that the conductors are sufficiently spaced out from each other to be able to disregard proximity effects. We can then suppose that the currents are uniformly distributed around the conductor and the periphery of the shielding. In other words, the magnetic field does not penetrate the conductors; the flow is thus calculated between the shielding and the conductors. To take the shielding into account, the self-inductance [3.23] and mutual inductance [3.24] are calculated using the images with the help of the transformations in keeping with [WEB XX], used in the particular case of rotational symmetry.

$$L_{ii} = \frac{\mu}{2\pi} \cdot ln\left(\frac{R_{shield}^2 - d_i^{\ 2}}{R_{shield} \cdot R_{C_i}}\right) \qquad [3.23]$$

$$L_{ij} = \frac{\mu}{2\pi} \cdot ln\left(\frac{d_j}{R_{shield}} \cdot \sqrt{\frac{(d_i \cdot d_j)^2 + R_{shield}^4 - 2 \cdot d_i \cdot d_j \cdot R_{shield}^2 \cdot \cos(\theta_{ij})}{(d_i \cdot d_j)^2 + d_j^{\ 4} - 2 \cdot d_i \cdot d_j^{\ 3} \cdot \cos(\theta_{ij})}}\right) \qquad [3.24]$$

For the dimensions provided earlier, the self- and mutual inductances are, respectively, calculated at 220 and 40.5 nH/m. These results give an acceptable order of magnitude with regard to those provided by the measurements (Table 3.3). The linear resistance terms (r_{ii}) can be calculated using the resistance expression defined in a DC state modified by the skin effect. Expression [3.25] is a simplification of the exact relation which normally includes Bessel functions of the first kind. Once the matrices of the resistive components r and inductive components L are determined, the capacitance and conductance can be deduced from the usual wire transmission relations, assuming the space is homogeneous ([3.26] and [3.27]) [PAU 92].

$$r_{ii} = \begin{cases} \frac{1}{\sigma_{Cu} \cdot \pi \cdot R_{C_i}^{\ 2}}, & R_{C_i} < 2 \cdot \delta \\ \frac{1}{2 \cdot R_{C_i}} \cdot \sqrt{\frac{\mu_0 \cdot f}{\sigma_{Cu} \cdot \pi}}, & R_{C_i} > 2 \cdot \delta \end{cases} \qquad [3.25]$$

$$[C] = \mu \cdot \varepsilon \cdot [L]^{-1} with\ \varepsilon = \varepsilon_0 \cdot \varepsilon_r\ et\ \mu = \mu_0 \qquad [3.26]$$

$$[G] = \frac{\sigma_{isol}}{\varepsilon} \cdot [C]^{-1} with \sigma_{isol} = conductivity\ of\ the\ insulating\ space \qquad [3.27]$$

The multiconductors transmission lines (MTL) method is based on the definition of differential equations which let us express the current and voltage of the conductors in terms of time and space. This system defined by a set of N conductors and a reference conductor which in our case corresponds to the shielding (Figure 3.31).

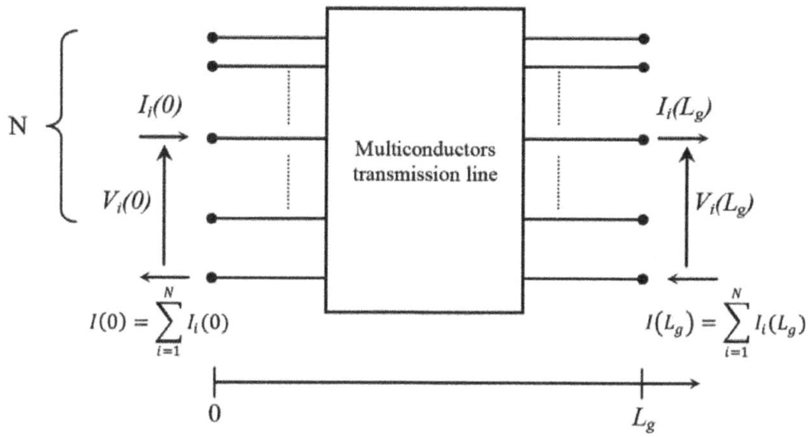

Figure 3.31. *Input and output values of a multiconductor transmission line*

This method [PAU 92] introduces the propagation matrix ϕ of the cable [3.28], which lets us define the electrical values at the input of the cable in terms of the output values for a given length (L_g). For example, for a shielded cable with four conductors, the dimensions of the matrix ϕ are [8 × 8] and those of the submatrices ϕ_{ij} are [4 × 4]. The resistive and inductive elements determine the Z matrix [3.29]. Similarly, the Y matrix is defined in terms of capacitive components and the conductances [3.30]. The submatrices of ϕ [3.34] are determined by calculating the determinant γ [3.31] of the product Y·Z and the transformation matrix T. We also need to define the characteristic impedance matrix Z_C [3.32] and its inverse Y_C [3.33].

$$\begin{pmatrix} V_{L_g} \\ I_{L_g} \end{pmatrix} = \begin{pmatrix} \phi_{11} & \phi_{12} \\ \phi_{21} & \phi_{22} \end{pmatrix} \cdot \begin{pmatrix} V_0 \\ I_0 \end{pmatrix} \qquad [3.28]$$

$$Z = r + j \cdot L \cdot \omega \qquad [3.29]$$

$$Y = G + j \cdot C \cdot \omega \qquad [3.30]$$

$$T^{-1} \cdot Y \cdot Z \cdot T = \gamma^2 \quad [3.31]$$

$$Z_C = Z \cdot T \cdot \gamma \cdot T^{-1} \quad [3.32]$$

$$Y_C = Z_C{}^{-1} \quad [3.33]$$

$$\begin{aligned} \phi_{11} &= Y^{-1} \cdot T \cdot \cosh\left(\gamma \cdot L_g\right) \cdot T^{-1} \cdot Y \\ \phi_{12} &= Z_C \cdot T \cdot \sinh\left(\gamma \cdot L_g\right) \cdot T^{-1} \\ \phi_{21} &= -T \cdot \sinh\left(\gamma \cdot L_g\right) \cdot T^{-1} \cdot Y_C \\ \phi_{21} &= T \cdot \cosh\left(\gamma \cdot L_g\right) \cdot T^{-1} \end{aligned} \quad [3.34]$$

Using a propagation matrix and its submatrices, it is possible to directly express the short-circuit impedances Z_{CC} and open-circuit impedances Z_{CO}, respectively, given by relations [3.35] and [3.36]. They are determined simply by setting $V_L = 0$ or $I_L = 0$.

$$Z_{CC} = V_0 \cdot I_0{}^{-1} = -\phi_{11}{}^{-1} \cdot \phi_{12} \quad [3.35]$$

$$Z_{CO} = V_0 \cdot I_0{}^{-1} = -\phi_{21}{}^{-1} \cdot \phi_{22} \quad [3.36]$$

We choose as an example the shielded cable with four conductors of 1.5 mm² introduced earlier (Figure 3.32). Its length is fixed at 5 m.

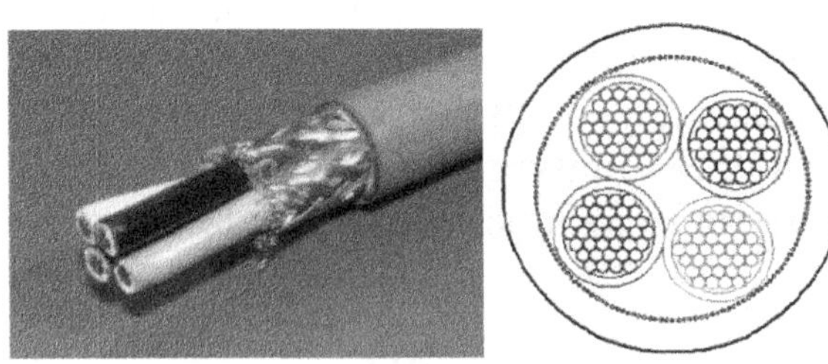

Figure 3.32. *Shielded cable 4 ×4.5 mm²/geometric model*

We can see that the behavior of the cable is correctly represented and that the resonance and antiresonance peaks are relatively well placed (Figure 3.33). Divergences appear, however, on the curves representing the impedance in open circuit for the lowest frequencies. These differences are nonetheless less pronounced on the short-circuit impedances as the resistive effects are dominant up to 10 kHz.

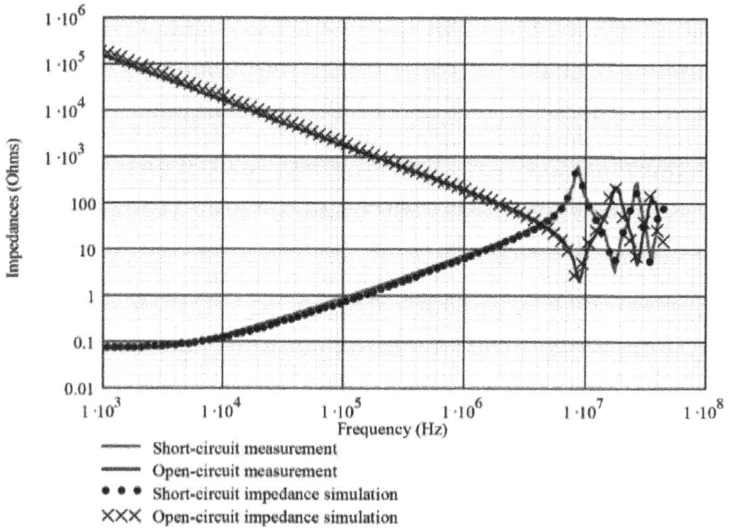

Figure 3.33. *Comparison between measurements and simulations on 5 m of cable. For a color version of this figure, see www.iste.co.uk/costa/electronics.zip*

3.5.2. *High-frequency model of an asynchronous machine*

We have seen the importance of a good representation of the propagation channels in the previous chapters, be it in common or differential mode. In the case of variable-speed drives, the electric machine defines a common-mode prioritized channel. The models considered to be relevant for an EMC study must at least cover the frequency range corresponding to the conducted emissions. Furthermore, the models introduced in this chapter require a strong assumption of machine linearity. This then suggests that the machine is never subject to saturations of the magnetic circuit. The components, therefore, only depend on the frequency, which lets us use the notion of impedance here.

3.5.2.1. *Simple model*

The usual models of electric machines are mainly developed to understand and describe their behavior for frequencies relative to rotational frequencies. These models are split into two groups. The first group includes the behavior in a steady state, established once the machine is determined as saturated or not (Figure 3.35). The second family lets us understand the behavior in a transient state of the machine in order to outline the coupling and/or speed control strategies. These models, although they have been outlined by the equations describing the electromagnetic interactions within the machine, cannot explain the behavior at high frequencies (>100 kHz). The geometric complexity and global heterogeneousness linked to the assembly of materials used (Figure 3.34) make it very difficult, and maybe impossible, to create an exact and direct analytical model. Therefore, it is necessary to approach this modeling step using behavioral models capable of rewriting the evolution of electrical quantities without delving too far into the detailed description of the physical phenomena.

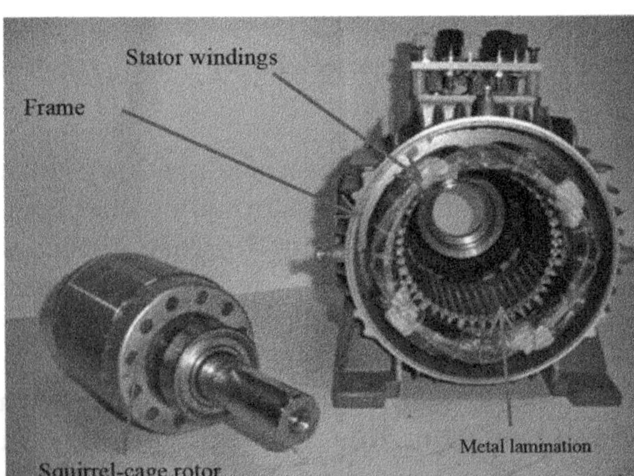

Figure 3.34. *Disassembled asynchronous machine*

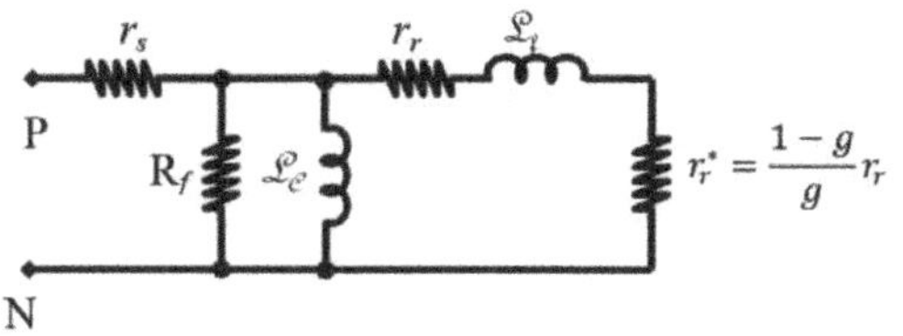

r_s:total resistance in series brought back to the stator (copper losses)
$\mathcal{L}_f$: Total leakage impedance brought back to the stator
$\mathcal{L}_c$: Cyclic inductance
r_r: Rotor resistance
g: sliding
R_f: Parallel resistance (iron losses through induced currents)

Figure 3.35. *Equivalent single-phase diagram*

With this standard model, the power of the machine is mainly defined in terms of the load on the tree modeled by the resistance (r_r^*) and the sliding (g).

The most common high-frequency asynchronous machine models are generally made up of a limited number of components like the one used in the previous chapter which clearly explains their frequency limit. Nonetheless, the simplicity of these models gives them the advantage of being able to determine the different components that they consist of with a minimal number of measurements. The initial outline of these models such as the one presented above (Figure 3.36) is generally based on an interpretation of physical phenomena that exist within the machine, and within a winding, to be more precise. In this specific case, the model is identical to that of a coiled inductance on a magnetic circuit, parasitic couplings of which will form with the frame. The self-inductance and the resistance of a winding are represented by Ld and Rs, respectively. The resistive component Rp can be linked with the losses induced in the iron lamination at the stator of the machine. Like the traditional model of an inductance on a coiled magnetic circuit, the component Cp represents the parasitic turn-to-turn capacitive couplings. The capacitive common-mode couplings between a winding and the frame of the motor are represented by the capacitance C_{g1} and C_{g2}.

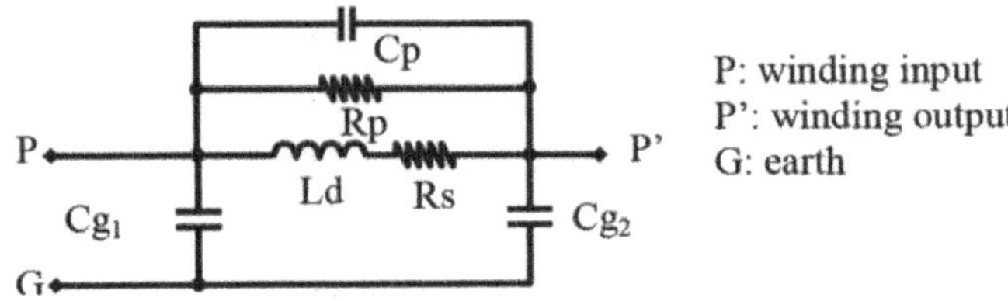

Figure 3.36. *Simple model of a winding*

The simple model presented above (Figure 3.36) can also be used to represent the impedance of a winding when the asynchronous machine is unloaded. In this case, the sliding is presumed to be null or sufficiently weak for the equivalent resistance to be supposedly infinite.

It is important to remember that this model is mainly intended for the common mode for frequencies distinctly greater than the frequency defined by the rotation speed. Therefore, it cannot take into account the operation of the machine's load, the equivalent model of which is recalled in Figure 3.35.

3.5.2.2. *Characterization: inductive parameters*

The model used in this section is outlined through an equivalent circuit for each phase in which the inductance L_d does not correspond exactly to the self-inductance value of a winding. Indeed, to determine this parameter, the three phases of the machine are connected in parallel. This approach thus allows the machine model to embed the mutual couplings between windings. The inductance L_d also takes into account the mutual inductances between windings. The equivalent impedance of a phase is deduced by multiplying the measured impedance by 3. Multiple measurements are, therefore, necessary. The first measurement consists of taking the impedances of three windings in parallel to directly determine the equivalent inductance of a winding (Figure 3.37). For this measurement, the earth potential (G) is not fixed, and the capacitance equivalent to the components C_{g1} and C_{g2}, both in series, is added to Cp. The measurements presented in this section are carried out on an AEG asynchronous motor of 750 W.

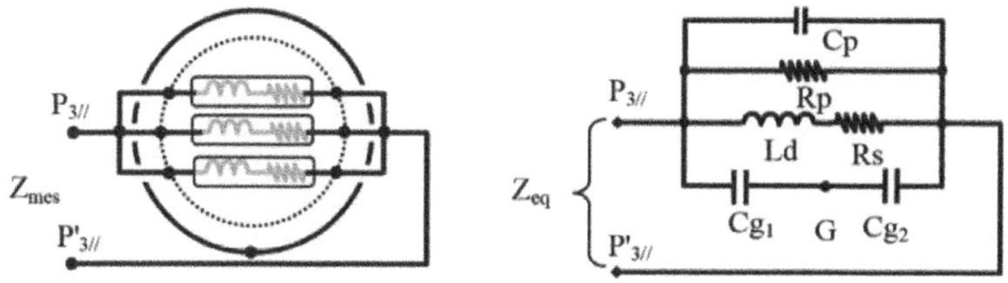

Figure 3.37. *Equivalent diagram of the impedance of a phase*

With this model, the impedance only shows one resonance frequency defined by L_d and the equivalent capacitance C_{eq} [3.37].

$$f_r = \frac{1}{2\pi\cdot\sqrt{L_d\cdot C_{eq}}} \qquad \text{with} \qquad C_{eq} = C_p + \frac{C_{g1}\cdot C_{g2}}{C_{g1}+C_{g2}} \qquad [3.37]$$

In this configuration, the impedance module measured at the impedance analyzer HP4194A is presented in Figure 3.38. The value of L_d is directly deduced from a point on the increasing section of this reading. At 1 kHz, the equivalent inductance is measured at 45 mH. The main resonance measured at 120 kHz (clearly visible in Figure 3.38) lets us make a first estimation of the equivalent capacitance (39 pF). The amplitude of the resonance peak is fixed by losses induced for the most part in the magnetic circuit. These are represented in the equivalent diagram as the resistance R_p.

The simulation resulting from these first results is not precise enough and differs from the measurements starting from the lowest frequencies (< 4 kHz). If the resonance peak seems to be correctly placed, the equivalent capacitance calculated previously is too weak since beyond the resonance the simulated impedance is much greater than the measured one. This error is essentially due to the simplicity of the model, which does not take into account the evolution of the winding inductance in terms of frequency.

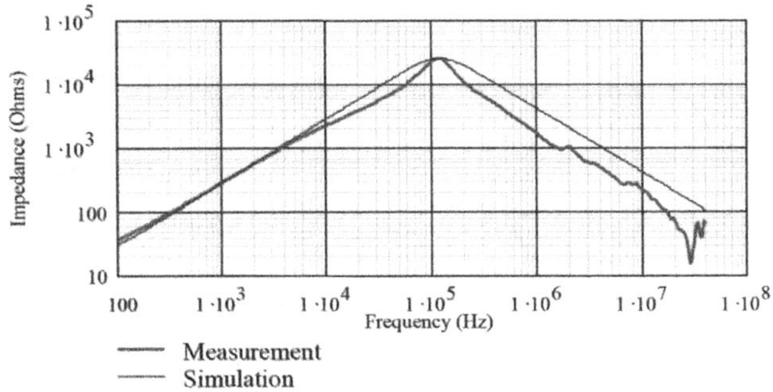

Figure 3.38. *Impedance of the three phases in parallel (×3)*

3.5.2.3. *Awareness of inductance variation*

The variation of the inductance is essentially due to skin effects and induced currents which limit the penetration of the field in the magnetic circuit and changes the current density in the conductors. These different combined effects cause a heavy decrease in the inductance in terms of frequency which can reach 10 mH per decade. To increase the realism of the simulation, we can improve the inductance model by proposing the diagram shown in Figure 3.40 which consists of a ladder of inductances and resistances. Therefore, the diagram lets us obtain an equivalent inductance dependent on the frequency, the evolution of which is comparable to the measured one (Figure 3.39).

By once again estimating the inductance as close as possible to the resonance, the equivalent capacitance is calculated at 81 pF, which gives us a good estimation of the frequencies up to 1 MHz. Beyond that, new resonances appear, which proves the existence of parasitic effects which we did not take into account when establishing this initial model (Figure 3.41). At this stage, the three capacitances of the model are unknown.

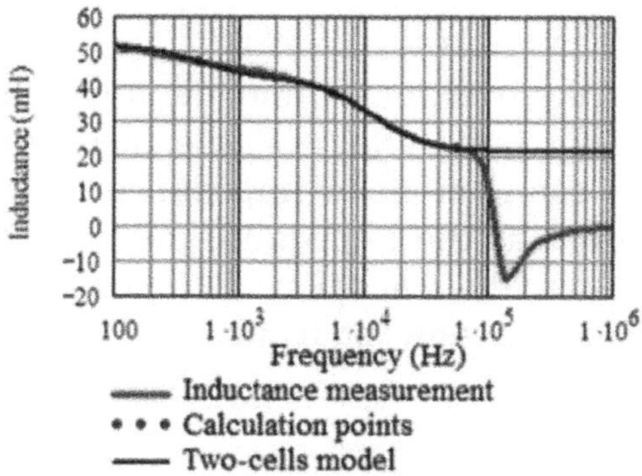

Figure 3.39. *Evolution of the inductance as a function of frequency. For a color version of this figure, see www.iste.co.uk / costa / electronics.zip*

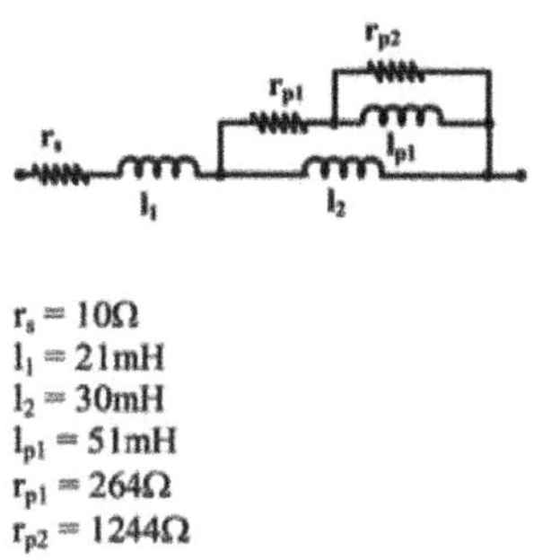

Figure 3.40. *Equivalent diagram of the impedance.*

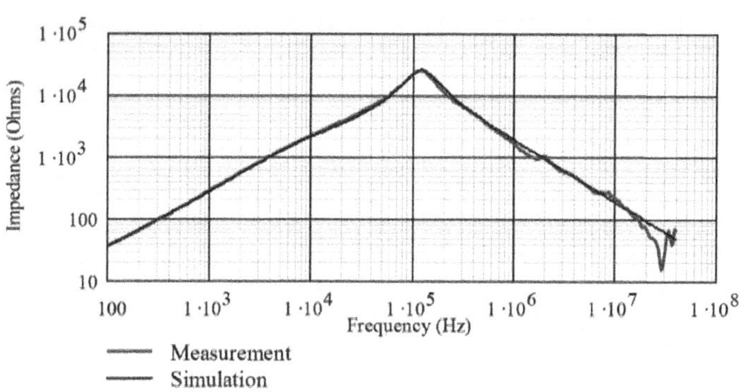

Figure 3.41. *Simulation / measurement comparison. For a color version of this figure, see www.iste.co.uk / costa / electronics.zip*

3.5.2.4. *Determination of the capacitance*

The second required measurement is carried out among the three windings in parallel with the earth (Figure 3.42). It represents the common-mode impedance of the motor. Its module, measured at the impedance analyzer and shown in Figure 3.43, shows the evolution of the capacitive couplings between the windings and the frame.

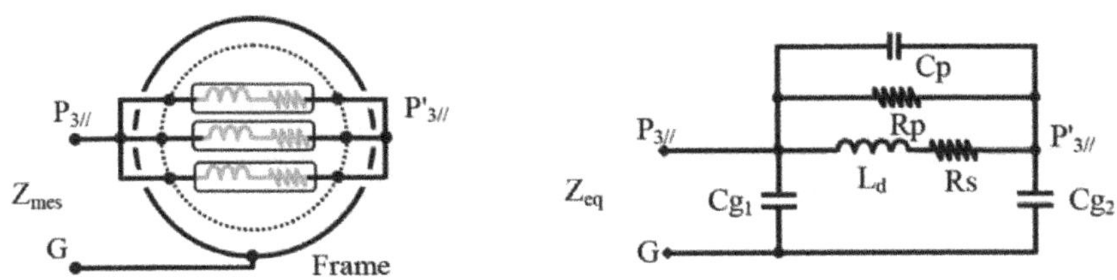

Figure 3.42. *Measuring conditions and equivalent diagram*

With this measurement, we can see that the biggest of the variations in capacitance occurs around the resonance frequency, identified in Figure 3.43 as 120 kHz. In this measured configuration, the inductive effects cause a resonance phase followed by an antiresonance. This particular evolution helps to determine the three capacitances of this model under certain presumptions.

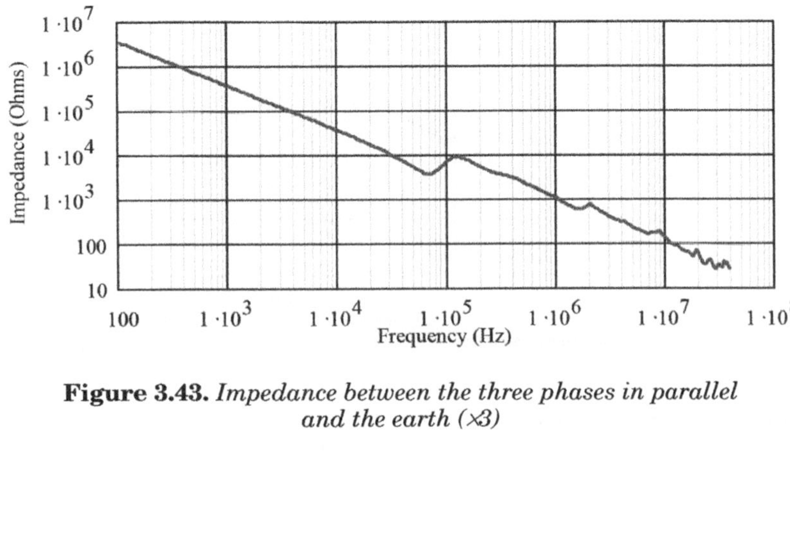

Figure 3.43. *Impedance between the three phases in parallel and the earth (×3)*

– At low-frequency, the impedance of the connection in series of L_d and R_s is negligible compared to that of C_p. Because of this, the capacitances C_{g1} and C_{g2} end up being in parallel and the equivalent capacitance is defined by the sum of these elements [3.38].

– At the highest frequency, the effects are reversed and it is the impedance of the inductance which becomes strong enough to be ignored. The equivalent capacitance, therefore, corresponds to the connection of C_p in parallel with C_{g1} and C_{g2} in series [3.39].

$$C_{low-frequency} = C_{g1} + C_{g2} \quad [3.38]$$

$$C_{high-frequency} = C_{g1} + \frac{C_p \cdot C_{g2}}{C_p + C_{g2}} \quad [3.39]$$

These two relations involving the equivalent capacitance defined by expression [3.37] help us obtain the results shown in Figure 3.44 due to an algorithm for the resolution of nonlinear systems.

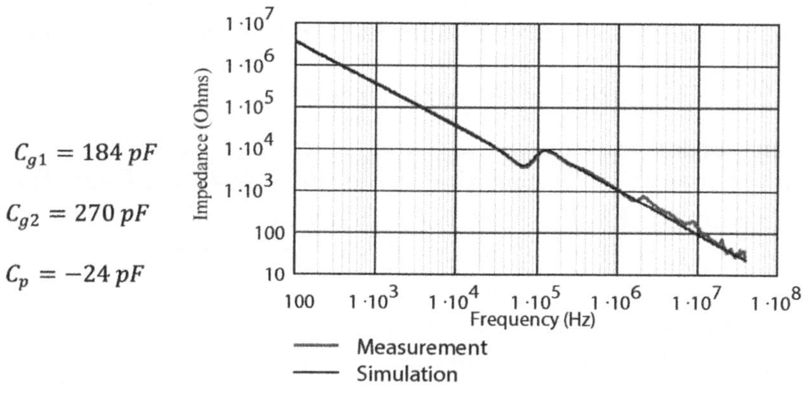

Figure 3.44. *Simulation / measurement comparison (earth motor phase impedance AEG750W). For color version of this figure, see www.iste.co.uk / costa / electronics.zip*

Regarding the capacitance, the results obtained are excellent for frequency below 1 MHz. The parallel resistance R_p estimated using the first measurement gives a satisfactory damping of the antiresonance at 120 kHz. To refine the resonance appearing at 70 kHz, a resistance in series with Cg_2 is required. The resulting global diagram is shown in Figure 3.45.

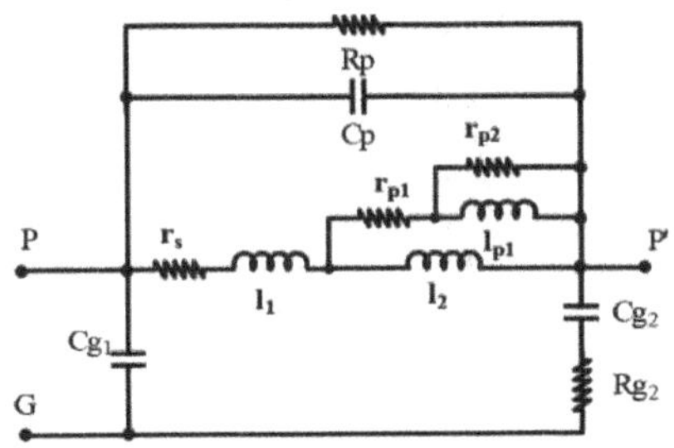

$C_{g1} = 184pF$ $l_1 = 21mH$
$C_{g2} = 270pF$ $l_2 = 30mH$
$C_p = -24\ pF$ $l_{p1} = 51mH$

$R_p = 30k\Omega$ $r_s = 10\Omega$
$R_{g2} = 1k\Omega$ $r_{p1} = 264\Omega$
$r_{p2} = 1244\Omega$

Figure 3.45. *Complete model*

The previously established assumptions for calculating the capacitance seem completely justified by the obtained simulation results. Nonetheless, we can verify them by determining these three elements in a more precise manner.

3.5.2.5. *Verification of the capacitance*

In order to separate and correctly determine the capacitances C_p, C_{g1} and C_{g2}, three measuring configurations are required. Each of these configurations allows us to only take two capacitances of the model into account at a time. With the help of appropriate short-circuits between the different ends of the motor, only two capacitances set in parallel are needed to define the frequency of the main resonance with the inductance Ld. This constraint helps us obtain the three measuring configurations shown in Table 3.4.

The capacitances C_{m1}, C_{m2} and C_{m3} come directly from the measuring of the impedance of these three configurations.

The system consisting of the three equations of this table lets us very simply obtain the capacitances we are looking for. The results obtained from these measurements are similar to the previous ones. Indeed, the capacitance C_P is unchanged as opposed to C_{g1} and C_{g2}, the sum of which is identical in both cases.

$$C_{g1} = 220pF \qquad C_{g2} = 233pF \qquad C_p = -24\,pF$$

<table>
<tr><th>Measurement between P and G</th><th>Measurement between P' and G</th><th>Measurement between P and G</th></tr>
<tr><td>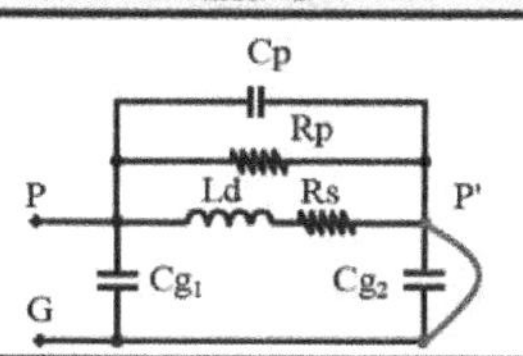
</td><td>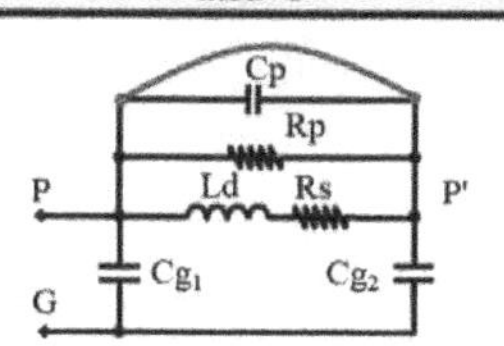
</td><td>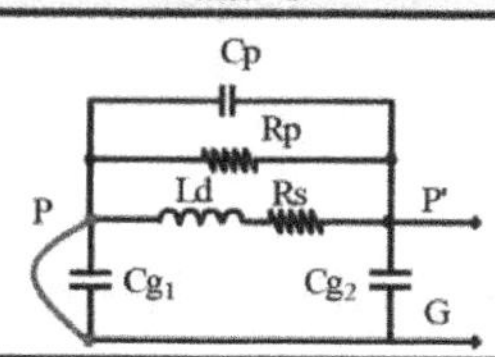
</td></tr>
<tr><td>$C_{m1} = C_{g1} + C_p$</td><td>$C_{m2} = C_{g1} + C_{g2}$</td><td>$C_{m2} = C_p + C_{g2}$</td></tr>
</table>

Table 3.4. *Measuring configuration corresponding to the model for extracting the capacitances*

NOTE.– Regardless of the calculation method, the capacitance C_p is negative. This type of result, which is common in modeling techniques, simply shows that a certain number of elements of the chosen model do not reflect reality. In our case, we have started with the representation and interpretation of the physical phenomena with local elements. We can now conclude that either the system of capacitances is incomplete or it is overdefined, and two elements are sufficient to obtain identical results. This diagram, therefore, outlines a behavioral machine and not a physical model. It is enough to model the behavior of the machine's frequency. Because of these different techniques, the equivalent generic model can be applied to other machines such as an asynchronous LEROY-SOMER motor 220 V 1,500 W 1,500 tr/mn. The equivalent impedance between a phase and the earth is shown in Figure 3.46.

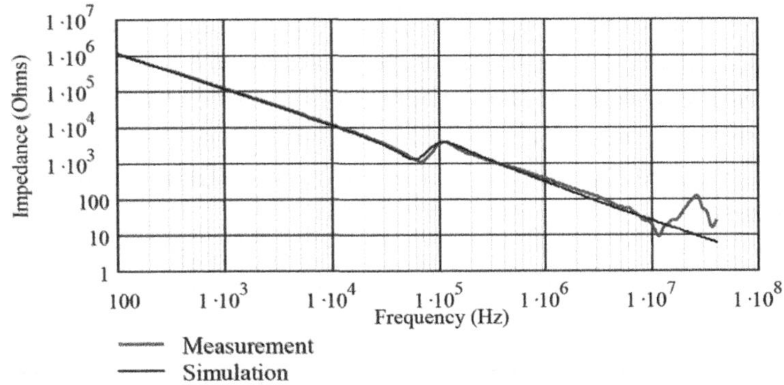

Figure 3.46. *Equivalent phase / earth impedance (LEROY-SOMER 1500W motor). For color version of this figure, see www.iste.co.uk / costa / electronics.zip*

3.6. Connection of the cable and the motor

3.6.1. *Total impedance read by the variable-speed drive*

We have studied and modeled the motor and its power cable separately. The next obvious step is to connect the two in order to define an entire model for the load. This step has the aim of bringing back the impedance of the motor to the output of the variable-speed drive through the power cable (Figure 3.47).

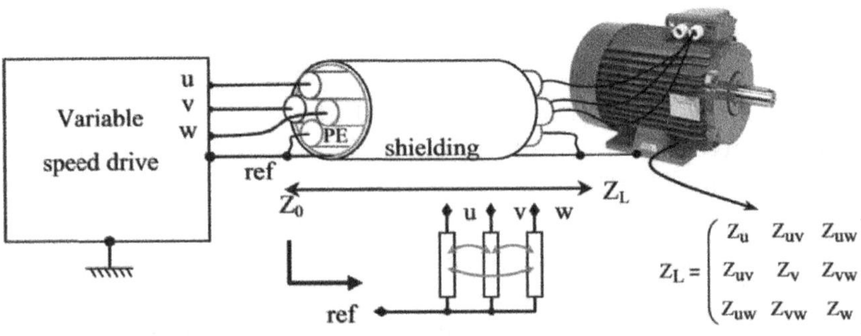

Figure 3.47. *Converter / cable / machine connection.*

The impedances linked to the motor help define the impedance matrix at the end of the cable (Z_L). This matrix, therefore, defines the voltages at the end of the motor in terms of current. Knowing the propagation matrix ϕ of the cable, the total impedance of the load seen by the inverter is expressed using expression [3.40].

$$Z_0 = (Z_L \cdot \phi_{21} - \phi_{11})^{-1} \cdot (\phi_{12} - Z_L \cdot \phi_{22}) \quad [3.40]$$

3.6.2. *Measuring the total common-mode impedance*

As we have shown in the beginning of this chapter, the machine constitutes a prioritized channel for common-mode currents. At low frequency, the total capacitance between the windings of the stator is around 1.5 nF for a small-sized motor (AEG750W). The shielded cable used here evidently causes an increase in the total capacitance between a phase and the earth [COS 12]. The impedance then decreases relatively quickly for sizable lengths (Figure 3.48). The impedance curve of the motor is translated and the resonance peaks are shifted toward the low frequencies. The resonances introduced by the components of the cable appear clearly for lengths greater than 1 m.

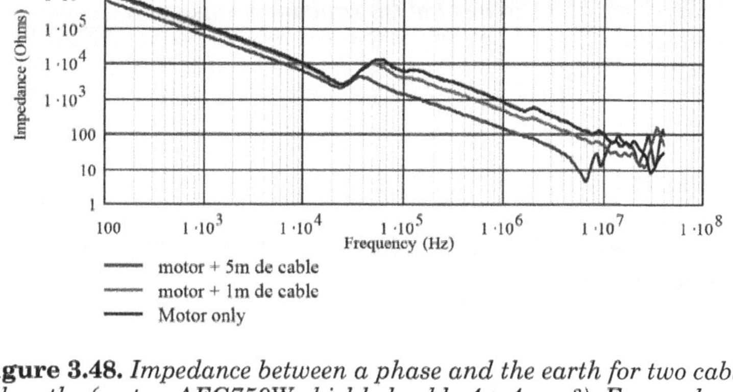

Figure 3.48. *Impedance between a phase and the earth for two cable lengths (motor: AEG750W shielded cable 4 ×4 mm²). For a color version of this figure, see www.iste.co.uk / costa / electronics.zip*

Figure 3.49 shows the evolution of the total capacitance of the load in terms of the cable length more clearly. For the motor on its own with the three phases coupled in star, we find the orders of magnitude of the results coming from the characterization: a high capacitance in low frequency and a weaker capacitance after the induced resonance through inductance of the windings. Regardless of the decrease of the parasitic capacitance of the cable with regard to the frequency, this will induce a quasi-constant increase ΔC of the equivalent capacitance over the entire studied frequency range.

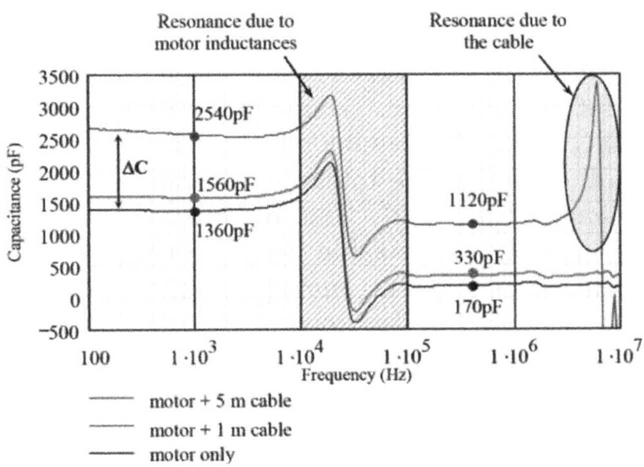

Figure 3.49. *Measurement of the equivalent capacitance between a phase and the earth for different cable lengths (motor: AEG750W shielded cable 4 ×4 mm²). For a color version of this figure, see www.iste.co.uk/costa/electronics.zip*

3.7. Results

3.7.1. *Time-based simulation and frequency-based simulation*

The first simulation shown below represents the estimation of the perturbations spectrum for a sinusoidal natural PWM, the modulation rate of which is fixed at 80%

in order to return to the previous simulation conditions. To validate this frequency-based approach and from this the simulation tool, the previous spectrum is compared to that of a time-based simulation using circuit-simulation software (Saber©). The aim of this study is not to verify the spectrum of a simulation based on another, but to show that a purely frequency-based approach, and the assumptions it requires, brings equivalent results in terms of calculation performance and a notably superior stability. The control law defined by this simulation is obviously the same as the one used for the frequency-based one. As we can see, the external envelope of the spectrum obtained through an FFT is identical to the one used in the proposed model (Figure 3.50). The differences that arise are the low level of the spectrum caused, on the one hand, by simply working with discrete signals and, on the other hand, by using fixed monitoring times. The use of dead-time becomes essential in order to ensure a realistic simulation of this type of structure, and also to limit the convergence problems caused by overly sudden variations in current during the sensitive commutation phases. The frequency simulation carried out using the model proposed in the previous sections gives similar results (Figure 3.51). Postprocessing is also carried out to present the spectrum that an analyzer with a "peak" detector would display.

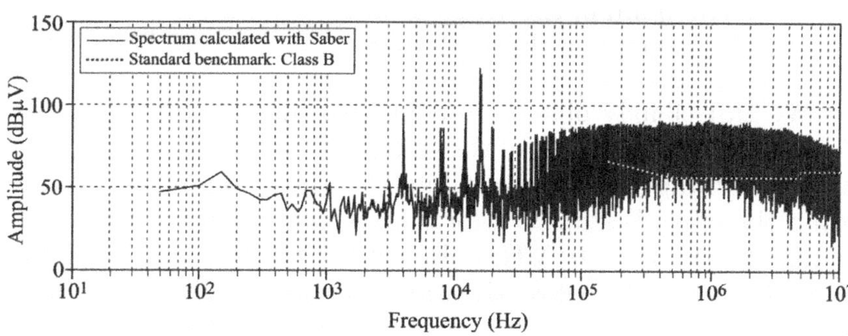

Figure 3.50. *Spectrum of the conducted perturbations calculated with SABER*

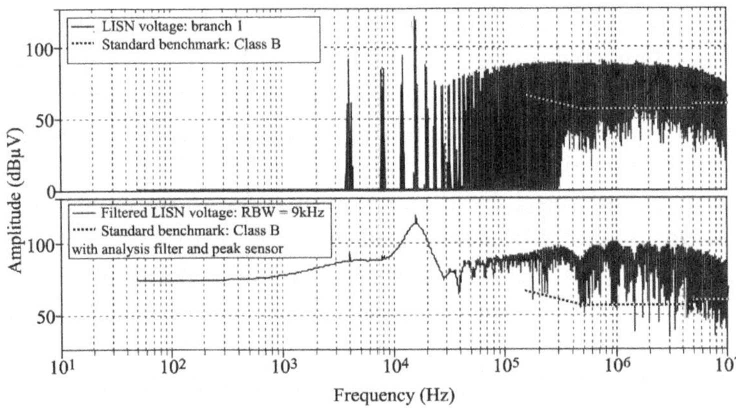

Figure 3.51. *Spectrum of conducted perturbations calculated with Matlab using the frequency method*

3.7.2. *Measurement versus simulation*

The test system of this study is an Altivar 28 variable-speed drive of 1.5 kW connected to a previously studied asynchronous machine (AEG 750W). For this test, the frequency of the machine is set to 40 Hz and the switch-mode frequency is required by the variable-speed drive to be at 4 kHz. At this frequency, the control law is a technique for the injection of harmonics with a limitation in the number of switchings (deadbanded) and a modulation ratio estimated to be 80%. The band of the video filter is set to 1 kHz in order to limit deformations in the spectrum.

We can clearly observe on the readings (Figure 3.52) the lobes from the filtering of the analyzer located at multiples of the switching frequency. Globally, the number is greater than the class B standard for a large section of the frequency range (150 kHz–10 MHz). Beyond 10 MHz, it diminishes quickly and falls below the class B profile. This significant decrease is, in part, due to the common-mode capacitive effects of the filter, which is always present in the structure. These initial measurements also show the impact of the motor since at 100 kHz we find a variation of the induced spectrum due to the decrease in parasitic capacitances between the phases and the earth. The results from the simulation are directly compared to the measurements (Figures 3.52 and 3.53). Therefore, we can note a good consistency between the simulation and measurements. Despite certain rays for frequencies greater than 1 MHz, the top level of the calculated spectrum, taking the filter and analyzer into account, is consistent with reality. These initial results help, on the one hand, to validate our model on a large frequency range and, on the other hand, to consider a study of the sensitivity of the main components taken into account in this example.

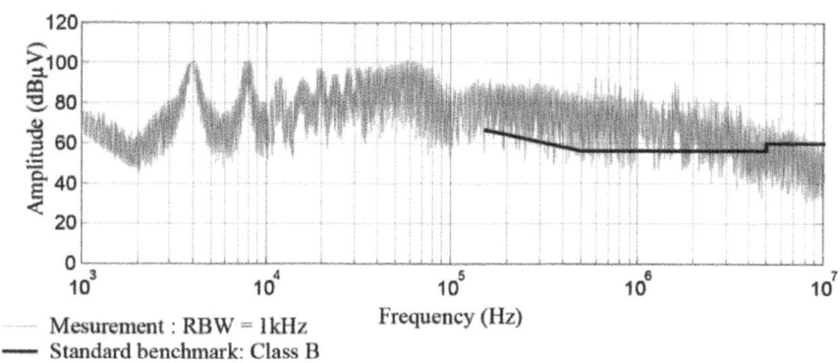

Figure 3.52. *Measurement of the conducted perturbations of the ATV28 variable-speed drive (1.5 kW) using an unloaded 1.5 kW LEROY-SOMER machine (branch 2)*

3.8. Passing from the time domain to the frequency domain: circuit simulations

The time-based simulation is an analysis and development tool, which has today become indispensable. It allows for a relatively simple verification of the functionality of the involved structure both on a control and on a power level. The abundance and the quality of the models of components developed by manufacturers and laboratories allow for an excellent degree of realism. However, these performances have a cost. Models of some active components such as IGBTs are very restrictive for simulation software. Incorporated into simulators such as SABER©, SIMPLORER© and PSPICE, these models are based on the solution of generally nonlinear equations relating to semiconductor materials. Because of this, they require long calculation time due to a large number of iterations required to ensure the convergence of the necessary electrical values. Furthermore, the complexity of the relations used often results in digital instabilities, which, in the worst-case scenario, stop the simulation.

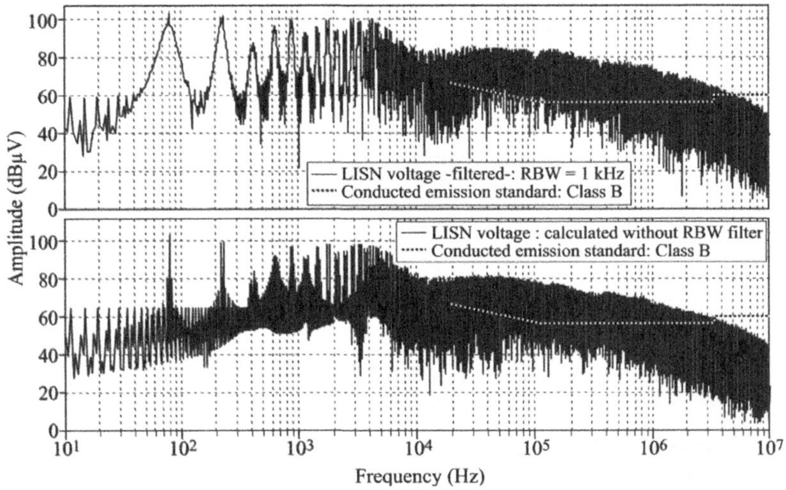

Figure 3.53. *Estimation of the spectrum of conducted perturbations on a branch of the LISN*

Specifically, the reason for time-based EMC simulations is to follow them up with an FFT in order to determine the representation of the spectrum of perturbing currents. The spectrum is then directly compared to the standards profiles. For a proper calculation of the spectrum, it is absolutely required that the processed signals be steady state. However, for complex structures such as those presented in this chapter incorporating components such as the LISN and/or the input filter, the use of transient states has increased considerably. The time constants required by different components naturally become very large compared to the calculation step. In theory, the latter must be less than the smallest time constant required in most cases by the parasitic components. These must inevitably be integrated into the simulation to take the perturbation phenomena into account. In practice, the calculation step (Δt) is chosen relative to the desired frequency range, while evidently respecting the Nyquist–Shannon criteria relevant to the sampling of signals [3.41].

$$\Delta t = \frac{T_{analysis}}{2 \cdot E(F_{max} \cdot T_{analysis})} \qquad [3.41]$$

with $E(x) = integer\ part\ of\ x$

The time window ($T_{analysis}$) over which the frequency study is carried out is not chosen randomly. Indeed, the realization of a spectrum study via Fourier transforms, whether quick or not (FFT/DFT), requires by definition that the processed signals be periodic [MAL 08]. Due to this, this window must evidently correspond to a whole multiple of the real period of the involved signal (Figure 3.54). In the opposite case, additional harmonics (or interharmonics) appear, which prevents the display of a correct image of the spectrum being considered (Figure 3.9).

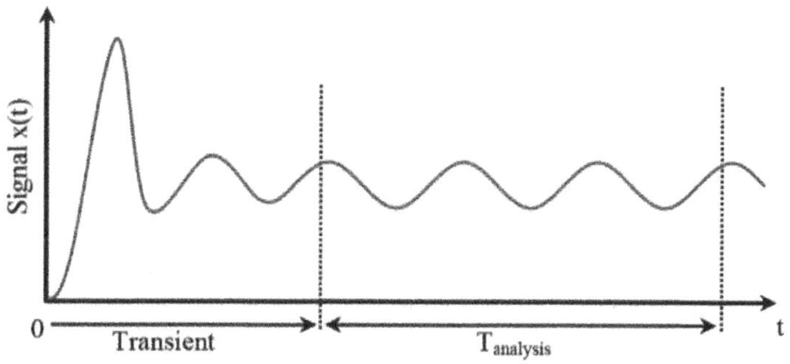

Figure 3.54. *Time signal: transient and steady states*

To summarize, the time-based simulation of the power structures carried out in an EMC study is extremely complex to implement. The convergence problems, linked to long calculation time, render the sensitivity studies delicate, with these being essential for the comprehension of the phenomena in play. On the other hand, the needs and digital resources become equally important requiring increasingly efficient materials (storage ability, main memory frequency of the processor, etc.).

3.9. Conclusion

The models proposed in this chapter allow us an adequate estimation of the conducted perturbations generated by a three-phase inverter used for the speed variation of an asynchronous machine. The two perturbation modes (common mode and differential mode) as well as the major propagation channels are represented. The cable and machine models help us know the perturbations in the load. Because of this, the phenomena of overvoltages at the ends of the motor cannot be estimated. The speed of this type of simulation also allows for the planning of optimization processes on the filtering components. Indeed, with an equivalent power calculation, the frequency simulations have helped to decrease the calculation time by 95%. The time-based simulations carried out with SABER©, which we have used for comparing the different estimation methods of spectrum estimation, offer less interesting performances in terms of calculation speed and especially robustness. As an example, the time-based simulation shown previously required 10 min of calculation time on an ordinary computer of 1 GHz with 512 MB of RAM. With a frequency-based model, the calculation time is less than 20 s for processing 260,000 frequencies.

3.10. Bibliography

[ADE 03] ADEME, "Les solutions pour optimiser vos systèmes motorisés", *Guide Technique*, Programme Motor Challenge, available at www.ademe.fr, 2013.

[CAC 99] CACCIATO M., CAVALLARO C., "Effects of connection cable length on conducted EMI in electric drives", IEMDC, 1999.

[CEN 04] *CENELEC, EN 61800-3:2004 Adjustable Speed Electrical Power Drive Systems - Part 3: EMC Requirements and Specific Test Methods*, 2004.

[CON 96] CONSOLI A., ORITI G., "Modeling and simulation of common mode currents in three phase inverter-fed motor drives", *Proceedings of EMC'96 ROMA, International Symposium on Electromagnetic Compatibility*, Rome, Italy, 17–20 September 1996.

[COS 12] COSTA F., GAUTIER C., REVOL B., *et al.*, "Modeling of the near-field electromagnetic radiation of power cables in automotives or aeronautics", *IEEE Transactions on Power Electronics,* vol. 28, no 99, p. 1, 2012.

[GUI 98] GUIRAUD J., Commande Vectorielle de machines alternatives à base de processeur de signal, PhD Thesis, INPG, February 1998.

[HOE 01] HOENE E., JOHN W., "Evaluation and prediction of conducted electromagnetic interference generated by high power density inverters", *European Conference on Power Electronics, EPE 2001*, Graz, Austria, 2001.

[MAL 08] MALLAT S., *A wavelet tour of signal processing: the sparse way,* 3rd ed. Elsevier Science & Technology, 2008.

[PAU 92] PAUL C.R., *Analysis of Multiconductor Transmission Lines*, John Wiley & Sons, New York, 1992.

[POP 99] POPESCU R., Vers de nouvelles méthodes de prédiction des performances CEM dans les convertisseurs d'électronique de puissance, PhD Thesis, INPG, December 1999.

[REV 11] REVOL B., ROUDET J., SCHANEN J.-L., *et al.*, *EMI Study of Three-Phase Inverter-Fed Motor Drives,* IEEE Transactions on Industry Applications, vol. 47, no 1, p. 223 231, February 2011.

[SEM 13] SEMIKRON, http://www.semikron.com/skcompub/en/miniskiip_ipm-93.htm, module Igbt MiniSkiip, 2013.

Chapter 4

Concrete Study of Solutions for the Reduction of Electromagnetic Perturbations

4.1. Concrete study of solutions for the reduction of electromagnetic perturbations

4.1.1. *Introduction*

The conformity to certain levels of electromagnetic fields required by the standards, both in terms of emission and immunity, forces manufacturers of electrical systems to implement specific filters and shielding. Indeed, even though everything is done during the design stage of a product to minimize electromagnetic compatibility (EMC) problems, the emission and susceptibility levels can be conformed to, with difficulty, without filtering or shielding. In a power converter, the size of the EMC filter can take up one-third of the total amount of space, and the shielding already puts a large number of constraints on the casings. The power cables and all of the interface cables are prioritized propagation channels for electromagnetic perturbations, and because of this, the filtering and the shielding must be studied together to ensure the best efficiency and to optimize the design.

To reduce the amount of emitted perturbations, the previous observations have shown that we can act on certain areas:

– Reducing the switching frequency F_o:

If, regardless of losses or dynamic operation of the converter, we were to have the choice of switching frequency, it would be preferable to choose the smallest value, as the graph shown in Figure 4.1 demonstrates. In these graphs, the maximum amplitudes of the current harmonics of the switch are shown for two switching frequency values. The decrease in switching frequency is demonstrated as a translation of the spectrum toward the lower frequencies.

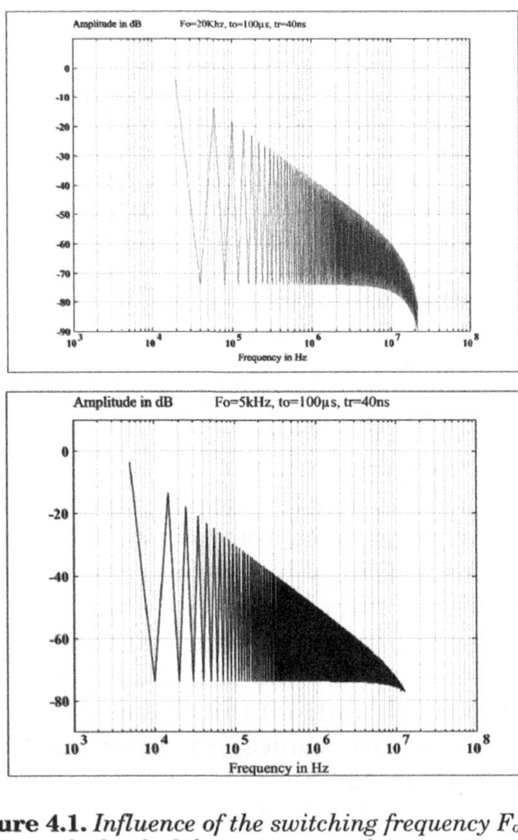

Figure 4.1. *Influence of the switching frequency F_o over the level of the common-mode current*

– Optimizing the choice of component's technologies:

These must incorporate the least parasitic components possible. We will note that the connections between these components introduce parasitic inductances proportional to their length (approximately 10 nH/cm). This aspect applies particularly to high-caliber semiconductors or power modules.

– Reducing $\frac{dV}{dt}$ or $\frac{dI}{dt}$, by slowing the control of the switch:

This solution can cause extra switching losses. This solution can be appropriate for converters operating at low frequency and is commonly used in variable-speed drives where we insert a gate resistor in order to reduce the switching speed of insulated gate bipolar transistor (IGBTs). A second solution consists of combining a soft switching structure with a slowed-down drive. For example, the zero-switching current (ZCS) mode enables a good control of dI/dt, and an appropriate gate drive allows us to slow down dV/dt during switching-on without a significant increase in losses [FER 99]. Figure 4.2 shows the obtained results on the common-mode current spectrum of a chopper in ZCS mode: the peak amplitudes are reduced by 15–20 dB between 10 MHz and 100 MHz.

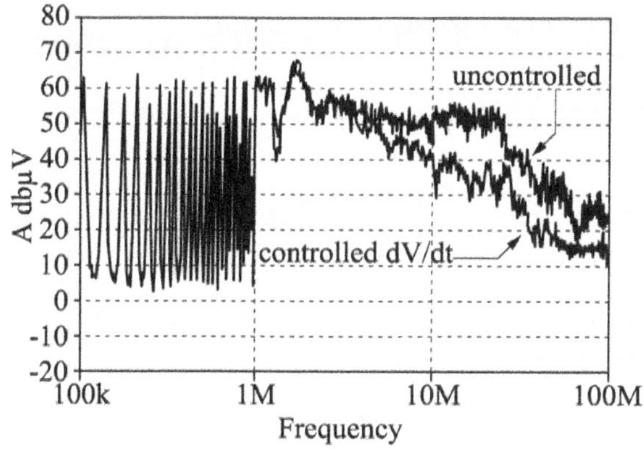

Figure 4.2. *Effect of the control of the switching transients*

– Reducing capacitive couplings (common mode):

This reduction can be achieved by using electrostatic screens under the switches or using shielding on sensitive cables as shown in Figure 4.3. The screen connection must be non-inductive for this device to be efficient [WAN 04].

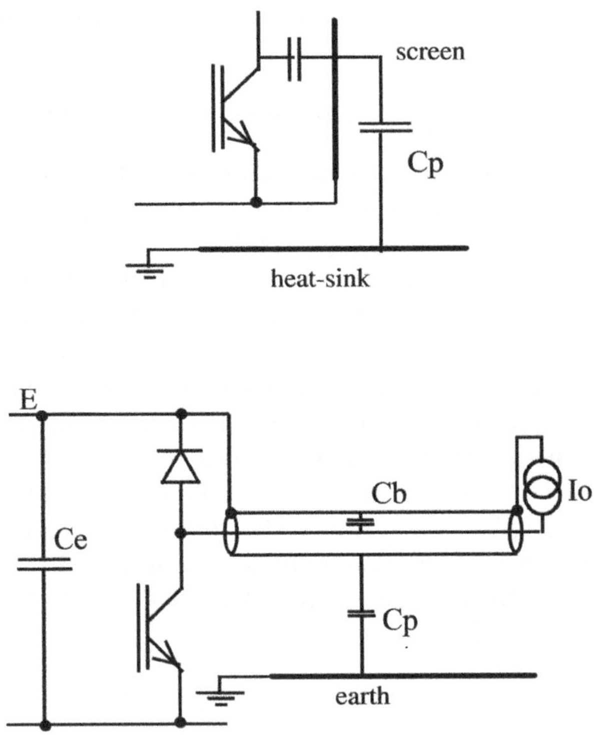

Figure 4.3. *Screens and shielding in the structure*

Furthermore, the cabling and/or routing can be handled in a way that reduces the surface of strongly varying potentials.

– Reducing the area of loops where the current varies strongly (differential mode):

The cabling must be organized in a way that minimizes the surface of radiating mesh. This reduces the parasitic inductance of the mesh and minimizes the radiating area [MAG 13].

– Reducing connection inductances:

For the same reasons, the organization of the cabling must be tidy and done in a way that minimizes the inductances in switching cells and the connection inductances of capacitors.

– Reducing the impedance of the ground plane and of the currents return path:

These last points effectively concern the quality of connections and the care taken in the creation of a converter. Minimizing the inductance requires that the area of the loop traversed by the current be at least the equivalent in two-dimensional (2D) connections of back-and-forth tracks in close proximity to each other. Busbar technology consists of laminating conductive plates separated by insulants and creating the necessary links and entrances; this is one of the possible industrial solutions used for optimizing cabling.

Figure 4.4. *Example of Busbar Rogers RO-LINX©*

In the following, we will pay particular attention to a design that manages to conform to the standards concerning conducted perturbations, through the study of EMC filters, and for radiated perturbations, through the study of shielding.

4.2. Filtering conducted emissions: analysis and conceptual design of common-mode filters

Nowadays, virtual prototyping has become essential in development phases and must evidently include the set of critical constraints involving the design of power converters (efficiency, cost, size, weight, heat, etc.). In this section, components involving the size of common-mode filters are discussed. These components, present in nearly every commercial power structure, are generally designed using the measurements of a prototype and a "trial and error" approach. Even though this process allows it in the majority of cases to comply with the constraints of standards, it is however not satisfactory and does not allow us to define components representing "the best solution" for the filtering of the developed structure.

4.2.1. *Introduction*

Even today, in power electronics, it is only once the prototype is completed that precertification tests are carried out, and that we properly calculate the EMC filter tasked with reducing the noise in conducted mode. Most of the time the EMC, be it conducted or radiated, is therefore not managed as a constraint during the design stages but as a test that the prototype undergoes before production. The study that we propose in this article is based on the modeling and design of EMC filters for electronic power converters such as variable-speed drives. Like nearly all of the electrical equipment intended for European market, variable-speed drives or converter/machine combinations must comply with the standards of conducted and radiated EMC. Even if their use, dominant in the industrial sector, would allow them to sometimes sidestep this constraint, their increasing use in the service sector requires them to fulfill or at least be prepared for all aspects of standardization. The modeling, characterization, design and optimization of filters are still a challenge for power electronics engineers. In general, these EMC filters are implemented using discrete components, which introduce certain limits. First, because of parasitic elements in these components, the efficient frequency range of the filtering is generally less than a few megahertz. Certain analyses presented in the following sections clearly show that the attenuation of filters is mainly determined by these parasitic elements. Before analyzing these second-order phenomena, it is essential to understand and master the operation of these very specific filtering devices.

The use of filters over a large frequency range is generally inevitable and even though commercial complete filters exist, it is generally preferable for cost reasons to design them directly in terms of the power structure being developed. There are many studies on Radio Frequency, (RF) filtering, proposing both passive and active solutions of interest [YOC 03, CAN 04]. In most cases, at least for conventional structures, the limitation of conducted emissions, which may be in common or differential mode, is achieved using simple structures with low-pass filters. The topology of the filter naturally depends on the type of perturbations (parasitic voltages or currents). We will now focus on the elementary filtering cell, which helps reduce high-frequency currents generated by a source in P_2 (Figure 4.5). To reduce the level in P_1, we must limit the propagation of currents by introducing a return path Z_Y with low impedance for high frequencies, particularly those defined by the standards [CEI 99]. The components that can ensure this function are capacitors. To ensure that the impedance required by these capacitors is lower than that of the grid, it is essential to add to the line, in series, a component with high impedance over a large frequency range (Z_X). This dual-function component is then inductive.

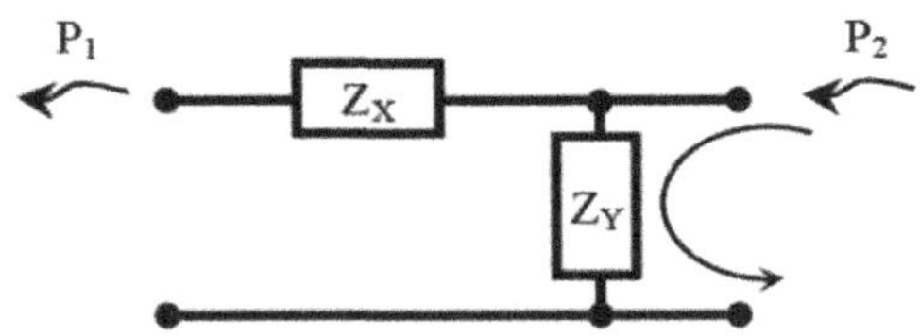

Figure 4.5. *Structure of a current filter*

4.2.2. *Description of a common-mode filter*

4.2.2.1. *Representation of the elements in common mode*

To understand the structure of a common-mode filter, it is preferable to introduce the ideal components into a single-phase structure. The common-mode currents propagate by definition on the two power lines to then loop via the earth; it is required that the two lines be decoupled (CYi). The simplest structure would have the inductance inserted into the earth line (Figure 4.6), however this solution is not compatible with the rules for the protection of persons and equipment. Indeed, the metallic parts and the casing, accessible to touch, must be directly earthed without an additional inserted component. Two inductances can be connected in series on the power lines, however the voltage drops that they could cause through differential-mode currents could be harmful to the good operation of the structure. Thus, these components must be able to act in common mode and remain transparent for differential-mode currents. The solution is simple since it involves coupling two inductances on a magnetic core.

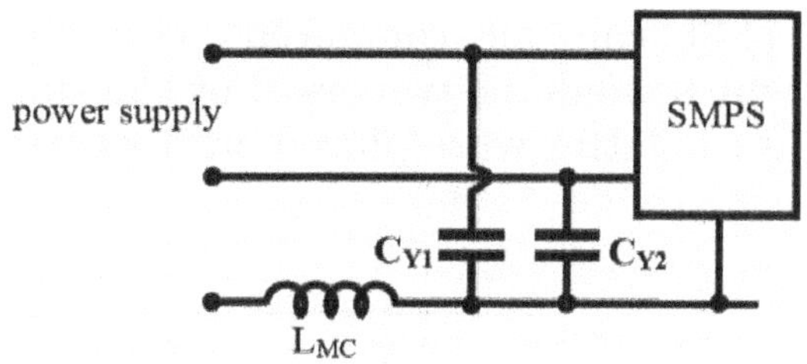

Figure 4.6. *Theoretical placement of the components in common mode*

4.2.2.2. *Common-mode inductance*

Some energy aspects allow us to quickly find an equivalent model with two perfectly coupled inductances (i.e. the coupling coefficient is supposedly unitary). The magnetic energy W_M stored in the inductances is given by relation [4.1]. For the two coupled inductances (Figure 4.7), the relations between the currents and the voltages are expressed simply using the matrix representation [4.2]. The magnetic energy that this element can store is given by equation [4.3].

$$W_M = \sum_{\{i,j\}} \frac{1}{2} \cdot L_{ij} \cdot i_i \cdot i_j \qquad [4.1]$$

$$\begin{pmatrix} v_1 \\ v_2 \end{pmatrix} = \begin{pmatrix} L_{11} & L_{12} \\ L_{12} & L_{22} \end{pmatrix} \cdot \frac{d}{dt} \begin{pmatrix} i_1 \\ i_2 \end{pmatrix} \qquad [4.2]$$

$$W_M = \frac{1}{2} \cdot L_{11} \cdot {i_1}^2 + \frac{1}{2} \cdot L_{22} \cdot {i_2}^2 + L_{12} \cdot i_1 \cdot i_2 \qquad [4.3]$$

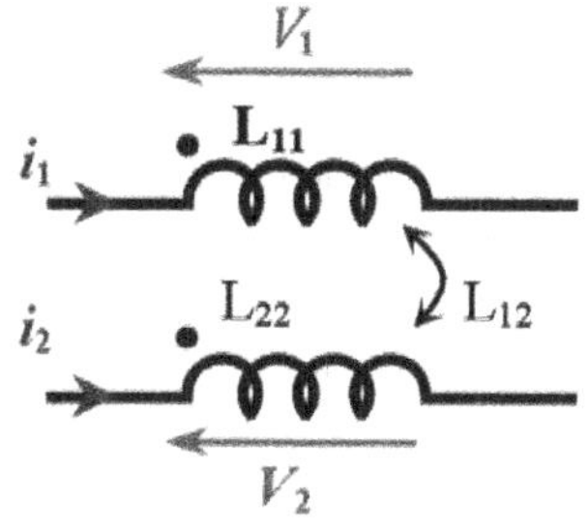

Figure 4.7. *Coupled inductances*

Relation [4.3] can be compared with that from an inductance, which would be traversed by the sum of currents (Figure 4.8) if the latter were placed on the earth connection [4.5].

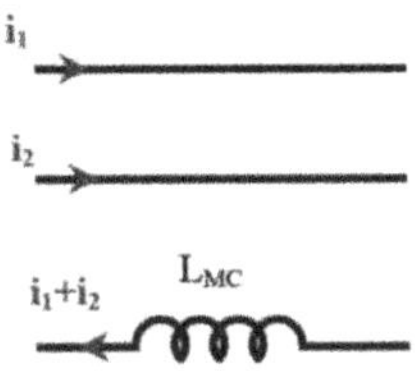

Figure 4.8. *Equivalent diagram of a common-mode inductance*

$$W_M = \frac{1}{2} \cdot L_{MC} \cdot (i_1 + i_2)^2 \qquad [4.4]$$

$$W_M = \frac{1}{2} \cdot L_{MC} \cdot {i_1}^2 + \frac{1}{2} \cdot L_{MC} \cdot {i_2}^2 + L_{MC} \cdot i_1 \cdot i_2 \qquad [4.5]$$

The identification of different terms shows that the two inductances coupled by a coefficient of a mutual positive inductance are strictly equivalent to an inductance traversed by the sum of currents, in other words, by the common-mode current. This element is, therefore, virtually inserted into the earth connection. The coupling is carried out in common mode, which is called common-mode inductance (CMI). This transformation proves that a perfect CMI is not defined only for the common-mode currents and remains invisible to differential-mode currents. Thus, the impedance Z_1 of the filter is defined by this component.

As we have seen previously, the use of a single passive element such as a CMI in a filter structure is generally not enough to sufficiently reduce the common-mode currents and thus comply with the standardized constraints. In most cases, a second-order filter is used to ensure the necessary reduction. Consequently, specific class Y capacitors are selected for common-mode filtering and placed between the different phases and the earth, usually closest to the perturbation source and far from the power supply. These filters are based on the assumption that the common-mode propagation channels are symmetrical. If this symmetry is not respected then there exists a common-mode conversion mechanism. Consequently, the capacitors and inductances must be identical (Figure 4.9).

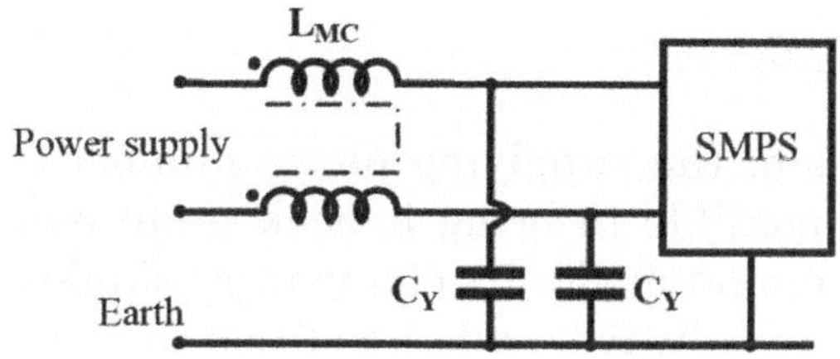

Figure 4.9. *Common-mode filter*

4.2.2.3. *Common-mode capacitor*

A similar study can be carried out for the capacitors C_Y. For this, we must define the common-mode voltage V_{MC}, its equation [4.6] being at the capacitor level (Figure 4.10).

$$V_{MC} = \frac{v_1+v_2}{2} \qquad [4.6]$$

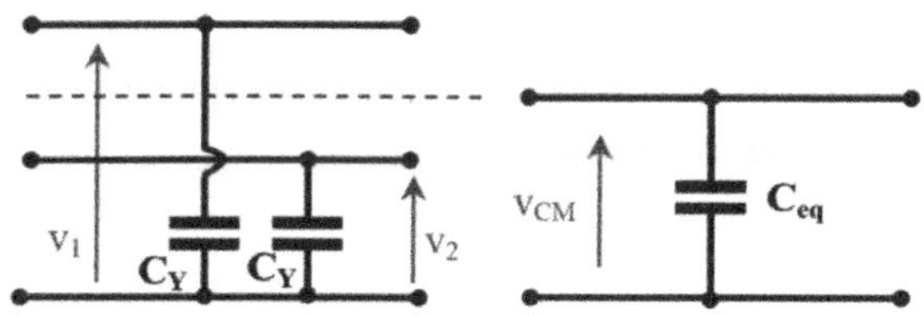

Figure 4.10. *Equivalent common-mode capacitor*

To obtain this in terms of an equivalent common-mode capacitor, the parallel currents should be written in each of these components [4.7] while including the common-mode voltage. Taking into account the symmetry of the electric power lines with the earth, the equivalent common-mode capacitor C_{eq} corresponds to the parallel connection of the capacitors C_Y [4.9].

$$i_{MC} = i_{Cy1} + i_{Cy2} \quad [4.7]$$

$$i_{MC} = C_Y \cdot \frac{dv_1}{dt} + C_Y \cdot \frac{dv_2}{dt} = 2 \cdot C_Y \cdot \frac{d}{dt}\left(\frac{v_1+v_2}{2}\right) \quad [4.8]$$

$$C_{eq} = 2 \cdot C_Y \quad [4.9]$$

Regardless of the supplying phase number of the system, it is always possible to bring it back to an equivalent filter defined in common mode. Its elementary single-cell structure (i.e. one level) is always made up of two components L_{MC} and C_{eq} (Figure 4.11). To determine the efficiency of this basic cell, the converter, generator of common-mode noise is modeled by a current source i_{MC}. The fictive network or line impedance stabilization network (LISN) is defined by Z_{LISN}. Following the same principle of base changing, the parallel connection of characteristic measuring impedances of the LISN equal to 50 Ω ($\forall f \in \{150$ kHz–30 MHz$\}$) lets us define an equivalent impedance Z_{LISN} in common mode, the value R_0 of which is assumed to be fixed at 25 Ω. The transfer function AF_{MC} between i_{MC} and i_0 [4.10], corresponding to the attenuation introduced by the filter, is defined by relation [4.11], in which Z_1 and Z_2 are, respectively, the ideal impedances of L_{MC} and C_{eq}.

$$AF_{MC} = \frac{i_0}{i_{MC}} \quad [4.10]$$

$$AF_{MC} = \frac{Z_2}{Z_{LISN}+Z_1+Z_2} \quad [4.11]$$

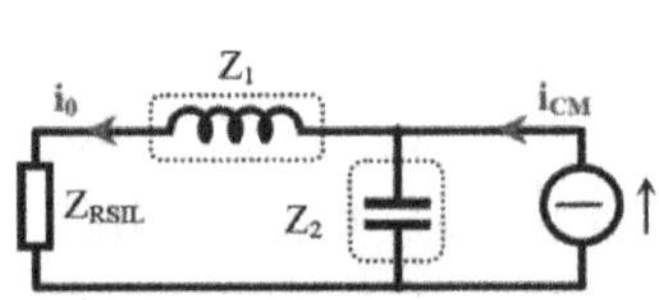

Figure 4.11. *Equivalent common-mode diagram*

4.2.2.4. *Asymptotic study*

A simple asymptotic study will help to determine the behavior of the filter. The presented diagrams assume that R_0 is less than the impedance of the capacitor with pulsation ω_0. This condition introduces a resonance to the pulse ω_1. If this condition is not met (R_0 is greater than Z_2 at ω_0), then there is no visible resonance on the attenuation curve.

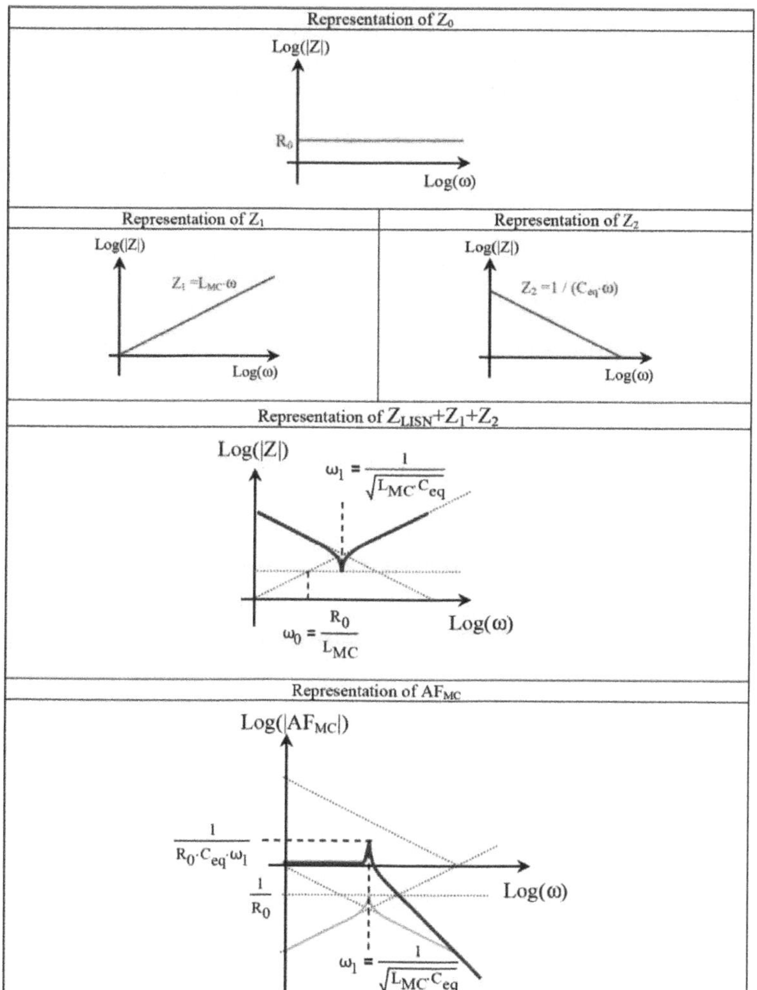

Figure 4.12. *Construction of the theoretical attenuation function*

A relatively simple approach lets us predetermine the value of the components of the filter. For this, we must determine the minimum attenuation necessary for drawing the spectrum of voltages measured on the LISN beneath the standardized benchmark required by the test equipment. As a general rule, this attenuation is found around the start of the imposed frequency range at 150 kHz (Figure 4.13). With the order of the filter N being required by its structure, we must determine the maximum level of the spectrum (S_M) given in dBμV at the frequency f_M. For this frequency, the limitation G_M (dBμV) is evidently known. The product $L_{MC} \cdot C_{eq}$ is, thus, given by [4.12]. This relation is deduced from the maximum cutoff frequency of the Nth-order filter, chosen in such a way that the required attenuation is achieved for the frequency f_M. For a simple design without optimization process, C_{eq} is initially chosen, complying with the security constraints required by the installation. For household appliances, differential protections of electrical equipment are sensitive to earth leakage currents. It is when the voltage is turned on that the inrush current in the filtering capacitors can surpass the detection limits of the circuit breakers. These common-mode capacitors must have a very high impedance at 50 Hz in order to ensure an acceptable level of zero current (<30 mA). Theoretically limited to a few ~100 nF, as a general rule, their value is voluntarily chosen to be less than 100 nF to guarantee a certain safety margin.

$$L_{MC} \cdot C_{eq} = \left(\frac{10^{\frac{S_M - G_M}{20 \cdot N}}}{2\pi \cdot f_M} \right)^2 \qquad [4.12]$$

with $N = 2 \times$ identical cell number.

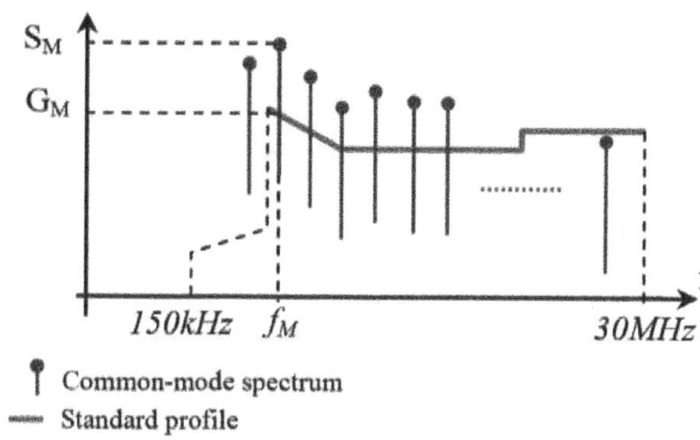

Figure 4.13. *Representation of the comparison between the conducted emissions spectrum and the standard profile*

4.3. Case study: determining a common-mode filter for a variable-speed drive

4.3.1. *Equivalent model of the drive*

To support this study, we will apply this filter research to a commercial converter, the input filter of which has been previously removed. This converter is a 5 kW variable-speed drive connected to a 3 kW asynchronous machine with a shielded three-phase cable of 5 m. The spectrum of common-mode perturbations produced by this equipment is measured at the terminals of an LISN. To keep this analysis independent of all of the measuring phases and to work only on common-mode emissions, the different parts of the tested combination (inverter, cable, AC motor, etc.) are measured and modeled to outline the equivalent EMC model of the inverter [LAB 10]. The LISN and the filtering cell are also incorporated into this model (Figure 4.14). The equivalent sources present in this model take into account the switched voltages and currents.

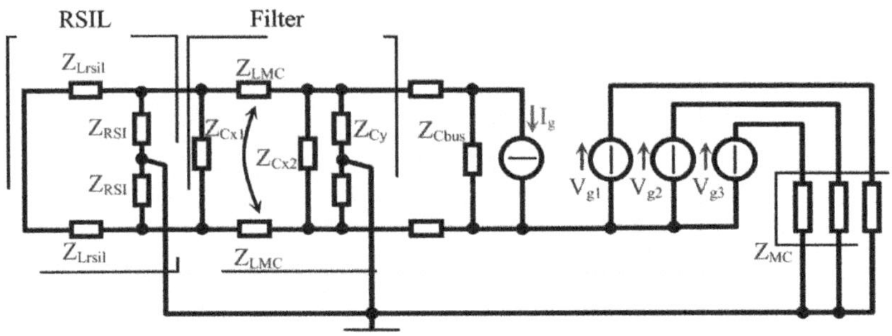

Figure 4.14. *EMC model of the variable-speed drive*

The common-mode currents are, therefore, calculated in the measuring branches of the LISN and let us determine the voltage spectrum to be compared with the profile similar to a direct measurement. The two spectrums being near-identical (Figure 4.15), we can validate the proposed model and use it to determine the impact of the filtering cell through simulation.

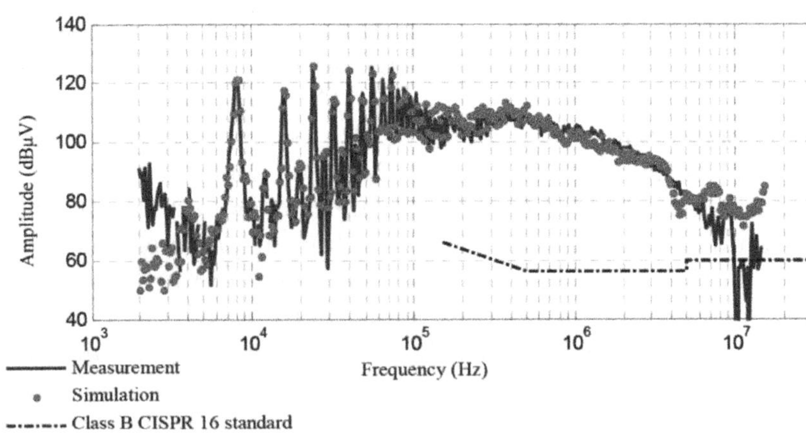

Figure 4.15. *Spectrum of the measured and simulated conducted emissions. For a color version of this figure, see www.iste.co.uk/costa/electronics.zip*

4.3.2. *Filter simulated using perfect components*

In the absence of the filter, the attenuation required to comply with the standard is near to 46 dB at 150 kHz. One of the local maxima of the spectrum is situated at 8 kHz and its level is of the order of 121 dBμV (Figure 4.16). This value highlights the switching frequency of the converter, generally set to be relatively low in order to limit, among other things, switching losses in the power switches. For this example, the value of the common-mode capacitors is set at 22 nF (<100 nF), thus 44 nF for the equivalent capacitor. The value of the CMI is given by equation [4.11]. The calculation gives a value of the order of 5 mH. This inductance is assumed to be perfect, in other words its value remains constant over the whole frequency range. The addition of these components to the model allows us to obtain a new spectrum that decreases quickly after the peak, which is linked to the resonance frequency of the filter set at 10 kHz. The effect of this resonance is clearly visible since the level of the spectrum at the switching frequency is 133 dBμV (Figure 4.17). This increase, linked to the interaction between the filter and the operating frequency of the converter, can modify the behavior of the CMI by causing its saturation. In the case of such an increase, the common-mode current in the inductance at the switching frequency is four times higher than the current determined without the filter.

Another solution consists of adapting the impedance of the filter to be a common-mode filter for the grid which, in our case, is fixed by the LISN. This adaptation removes the resonance of the filter. The condition required to ensure this adaptation is given by relation [4.13]. The couple of values $\{L_{MC}, C_{eq}\}$ is, therefore, required.

$$\frac{L_{MC}}{C_{eq}} = {R_0}^2 \qquad [4.13]$$

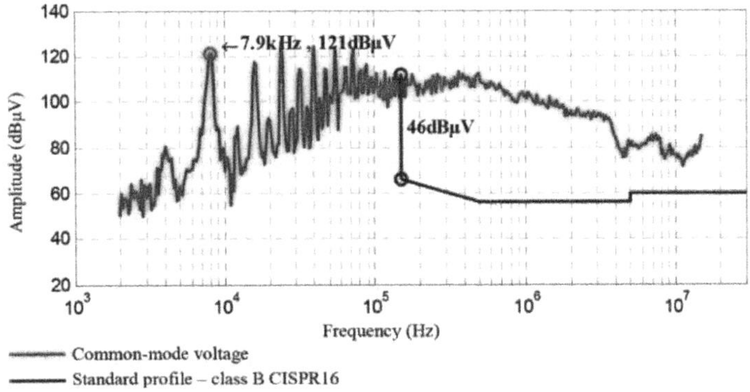

Figure 4.16. *Comparison of the spectrum with the standard. For a color version of this figure, see www.iste.co.uk/costa/electronics.zip*

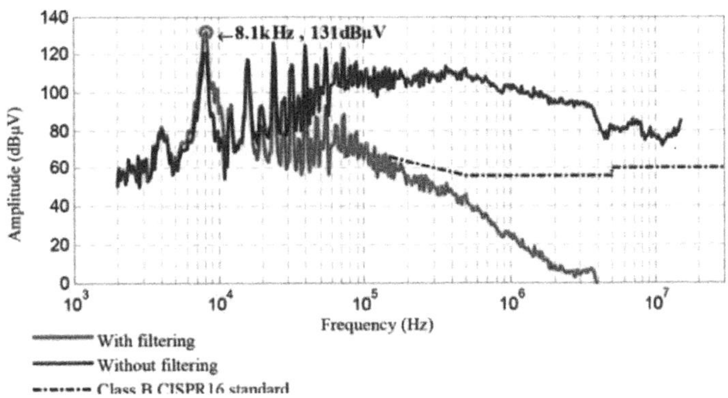

Figure 4.17. *Filtered and non-filtered spectrums. For a color version of this figure, see www.iste.co.uk/costa/electronics.zip*

In this case, the maximum level of the switching frequency remains approximately the same and the spectrum is correctly reduced (Figure 4.19). Unfortunately the components L_{MC} and C_{eq} are considerably modified since the obtained values are fixed at 380 μH and 300 nF, respectively. Even though the value L_{MC} is particularly interesting, the obtained value for C_Y (150 nF) is not advised for low-power applications (differential protection). Fundamentally, the impedance value of the grid modified by the fictive network is low, which leads to a high value of C at low frequencies. Evidently, in order to comply with the switching frequency of the filter, the chosen inductance value is small. There are two direct advantages from using small inductance values. The first advantage is a decrease in volume since it is determined in terms of the value of the inductance. The second advantage is related to parasitic elements, which depend on geometric characteristics. If we take a small inductance value, the number of winding turns is low and the magnetic cores are small.

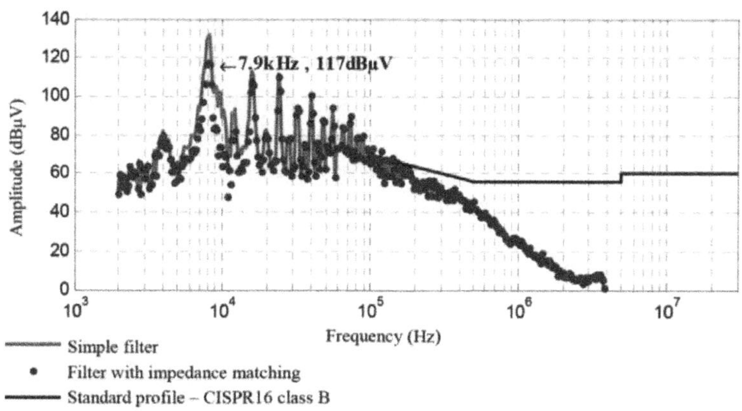

Figure 4.18. *Attenuation curve of the filter with and without impedance matching. For a color version of this figure, see www.iste.co.uk / costa / electronics.zip*

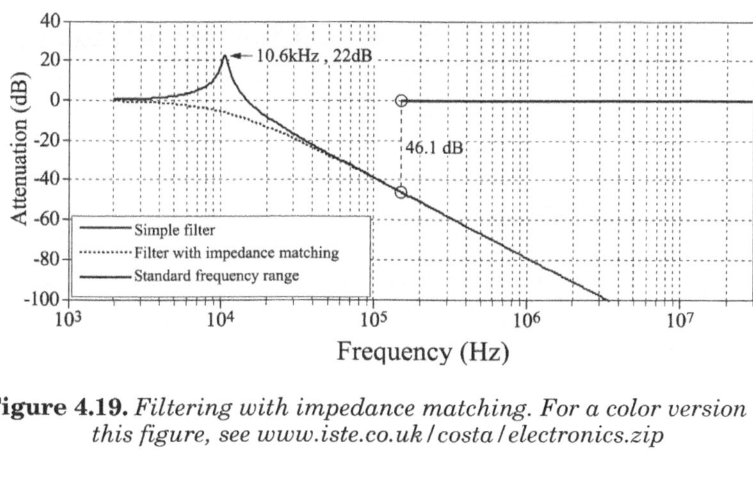

Figure 4.19. *Filtering with impedance matching. For a color version of this figure, see www.iste.co.uk / costa / electronics.zip*

4.3.3. *Effect of the parasitic elements of components*

The components of these filters must "theoretically" comply with two demands: they must tolerate in complete safety the operating voltage and the rated current of the electrical equipment without their high-frequency characteristics changing. In order to design these filtering cells, it is essential to understand, or even predetermine, the HF behavior of the components that constitute them. In other words, the components (inductances, capacitors) and connections cannot be simply considered as ideal over the frequency range used by the filter (hypothetically 150 kHz–30 MHz). It is always interesting to note that the duality relations are particularly applicable to frequencies such that the parasitic elements define the impedances of the components. Thus, a coiled inductance presents a capacitive behavior in HF whereas that of a capacitor is inductive [GRA 99, MAS 97].

The same approach as the one used to describe the attenuation curve of the perfect filter is applied to imperfect components. Each equivalent model of L_{MC} and C_{eq} is simple but realistic. The capacitors are modeled by a series connection of C, equivalent series inductance (ESL) and equivalent series resistance (ESR). The inductance can be simply defined as a parallel connection of components (L_{MC}, C_P and R_P). The asymptotic construction of the attenuation curve is a bit more difficult than the previous one, but takes the impact of parasitic elements into account very well (Figure 4.20).

Taking into account the parasitic elements of each component, the attenuation of the filter is limited in its frequency [WAN 04] and loses its distinctive attenuation of 40 dB per decade (Figure 4.21). Statistically, passive filters with components that are often of the same order of magnitude for differing equipment lose their efficiency for frequencies situated between 100 kHz and 1 MHz. The parasitic elements, for which predetermination is very difficult, also have typical orders of magnitude. The impact on the spectrum is obvious (Figure 4.22).

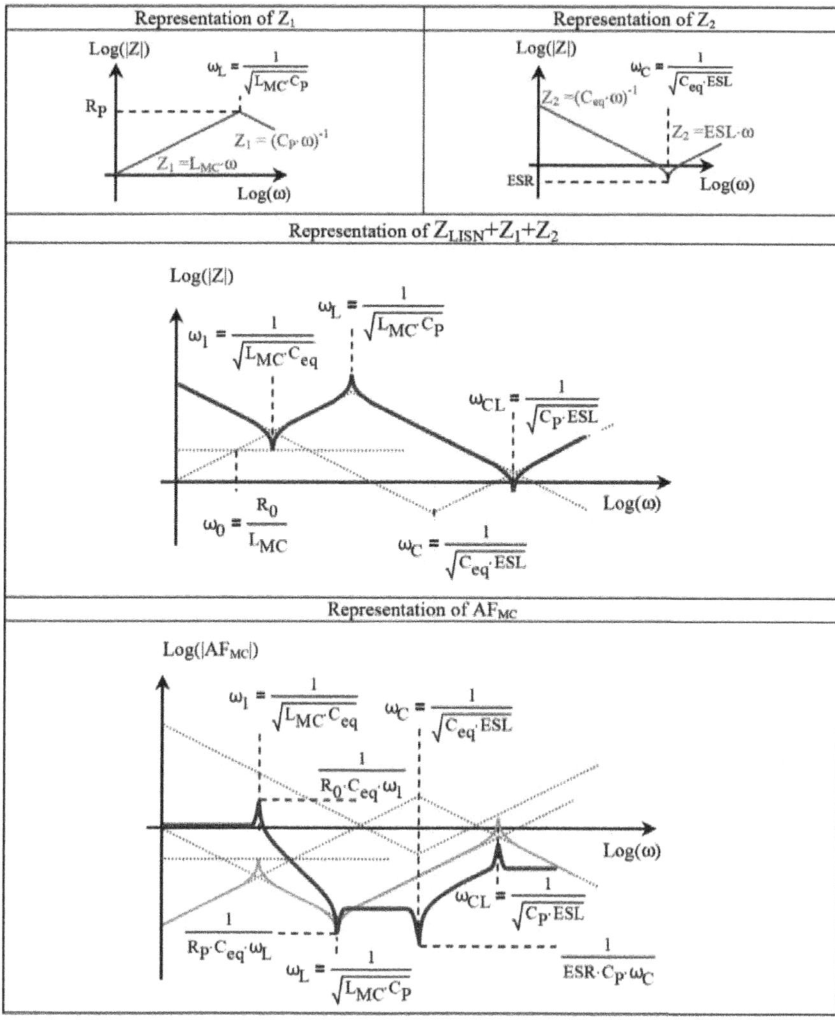

Figure 4.20. *Construction of the attenuation function with parasitic elements*

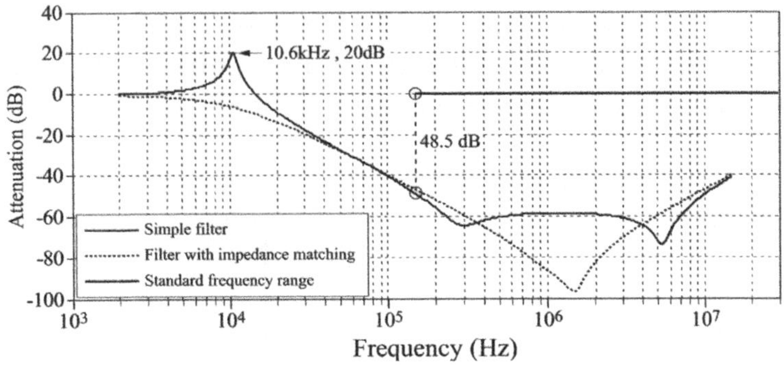

Figure 4.21. *Effect of parasitic elements on the attenuation curve of the filter with and without impedance matching. For a color version of this figure, see www.iste.co.uk/costa/electronics.zip*

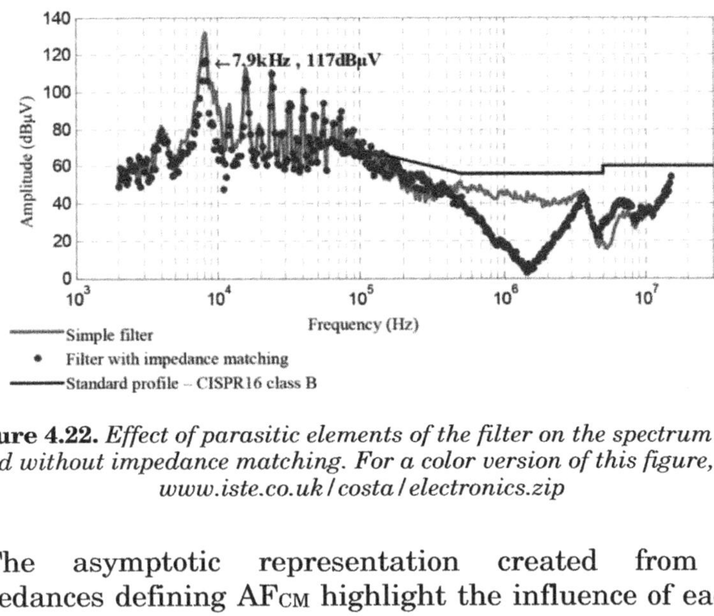

Figure 4.22. *Effect of parasitic elements of the filter on the spectrum with and without impedance matching. For a color version of this figure, see www.iste.co.uk/costa/electronics.zip*

The asymptotic representation created from the impedances defining AF_{CM} highlight the influence of each of the parasitic elements. To verify it, a measurement of the attenuation on a real filter (Figure 4.23) placed on a commercial variable-speed drive clearly shows the influence of these undesirable elements (Figure 4.24). From this curve, we can identify the resonance frequency of this filter, the expression of which is given by [4.12]. Contrary to the asymptotic representation, the real curve does not have a floor defined by the effects of the equivalent parasitic capacitance parallel to L_{CM} and the common-mode capacitors C_Y. This effect, which indirectly favors the efficiency of this filter, is linked to the conditions of two phenomena. The first can be simply explained by a low parasitic capacitance value C_P. For the filter in this example, the number of winding turns is relatively low for a toric magnetic core, the circumference of which allows for the creation of single-layer loops, thus limiting parasitic capacitive couplings. We must also take into account the decrease in permeability of the magnetic material with the frequency (Figure 4.25). This evolution naturally leads to a reduction of the inductance value repelling the self-resonance frequency of this element.

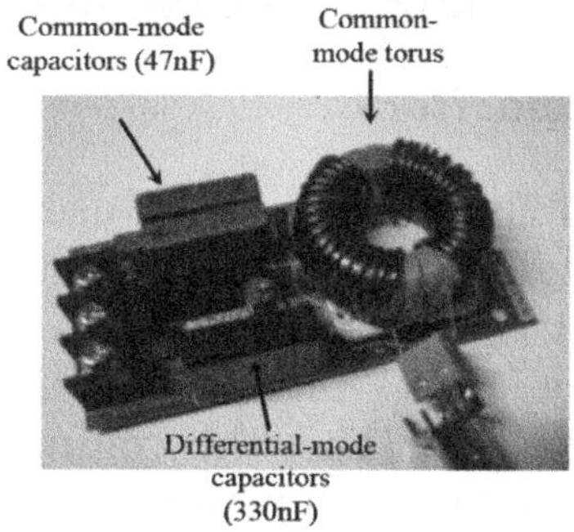

Figure 4.23. *Tested common-mode filter*

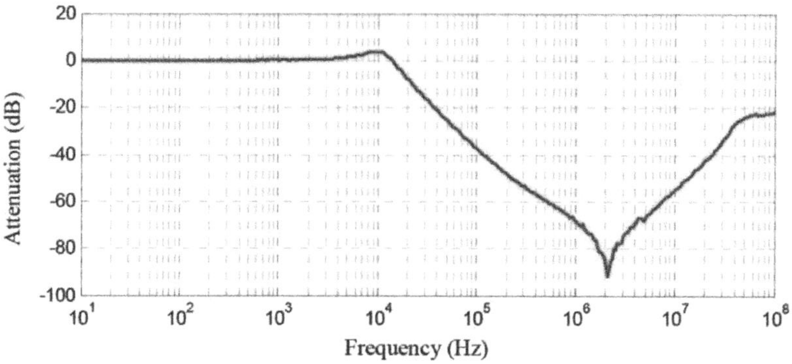

Figure 4.24. *Measurement of the attenuation of the filter*

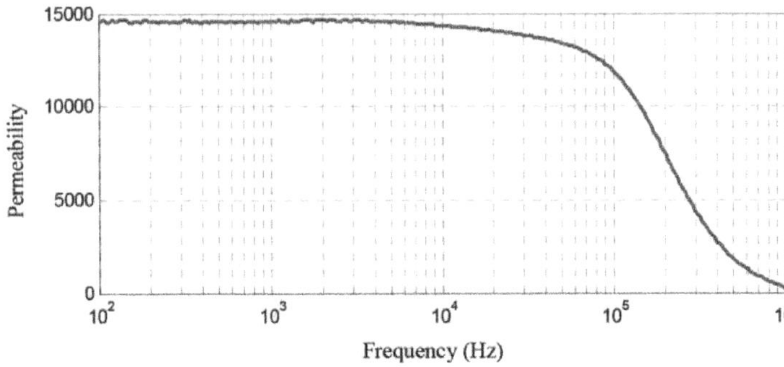

Figure 4.25. *Relative permeability of the common -mode magnetic ring core*

4.4. Design and optimization components

4.4.1. *Study of capacitors*

The majority of class Y capacitors used for common-mode filtering are created with a dielectric material of polyester (MKT) the permittivity (ε_{MKT}) of which is 2.7 for an electrical strength (E_{MKT}) of 10 kV/mm. From these data and the maximum usage voltage (U_m), the volume can be expressed as a linear capacitance function [4.14]. This common relation, verified for this type of technology, lets us predetermine the desired optimal volume of the capacitor of the filter (Figure 4.26).

$$Vol_{CY}(c) = \frac{c}{\varepsilon_{MKT}} \cdot \left(\frac{U_m}{E_{m_{MKT}}}\right)^2 + V_{0_{MKT}} \qquad [4.14]$$

We can also predetermine the parasitic components of these capacitors in order to best approach the response of the filter in the process of finding the most compatible components. The model relating to these capacitors is an R-L-C circuit in series, as we have defined it previously. Readings at the impedance analyzer HP4194A allow us to precisely determine these different parameters. Table 4.1 presents these components for a set of capacitors.

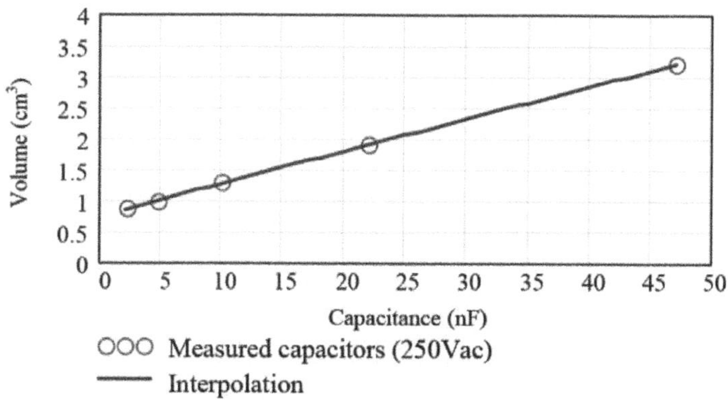

Figure 4.26. *Volume of the class Y capacitors*

Capacitance (nF)	1	2.2	10	22	47
Inductance in series (nH)	6.9	7.1	6.88	7.5	7.33
Resistance in series (mΩ)	444	383	240	113	90
Resonance frequency (MHz)	60.5	40.2	19.2	12.4	8.57

Table 4.1. *Parasitic components and resonance frequencies of the studied Y capacitors (MKT 250 V)*

We can see that the inductance in series does not vary significantly depending on the capacitance, contrary to the resistance in series (ESR), which rapidly decreases (Figure 4.27). We can, therefore, determine a relation that allows us to estimate this parameter in terms of the capacitance [4.15]. This resistance determines one of the maximum attenuations of the filter (Figure 4.20).

$$esr(C) = \alpha \cdot e^{-\beta \cdot C} + \gamma \qquad [4.15]$$

with $\alpha = 0.4$, $\beta = 1.1 \cdot 10^8$ and $\gamma = 0.08$.

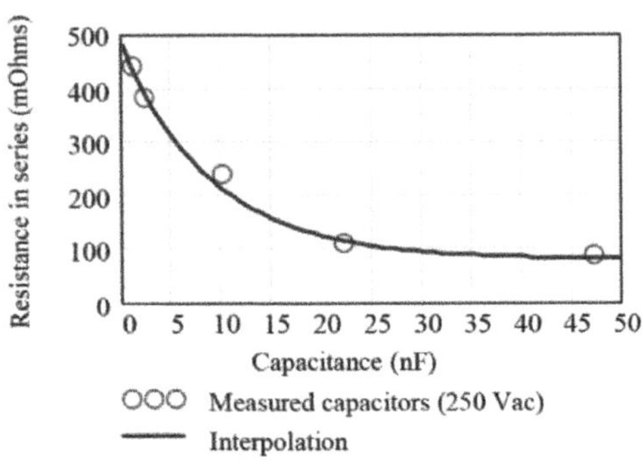

Figure 4.27. *Evolution of the resistance in series of the filtering capacitors*

4.4.2. *Study of the common-mode toric inductance*

4.4.2.1. *General thoughts*

The process proposed for realizing this type of inductance is based on the use of magnetic toric circuits (Figure 4.28) around which only a layer of conductors is used. The magnetic materials used for the common-mode filtering have high permeability. These materials are divided into two distinctive groups. The first and more traditional group is composed of ferrite-based materials with high permeability, whose saturation induction remains modest (<400 mT). The second group, associated with more recent materials, is composed of nanocrystalline alloys. These two types of competitively priced materials have very different characteristics. The second category, for example, offers performances that are, in terms of relative permeability (>25,000) and saturation induction (1.2 T), very interesting for high-frequency filtering. Nevertheless, the difficulty of conditioning nanocrystalline materials and their consistently high prices render their use marginal. Manganese–zinc-based materials, however, can also reach relatively high permeability ($\mu_r > 2 \times 10^4$) (example: the material 3E9 of FerroxCube) [MAG 13]. This relative permeability remains relatively constant over a frequency range, which can reach up to 100 kHz (e.g. 3E9 material, μ_r = 15,000 at 100 kHz). Beyond this frequency, the magnetic properties of the material decrease rapidly (–27 dB/decade). In the frame of this study, the retained magnetic circuits are ferrite based, the relative permeability of which is greater than 5×10^3.

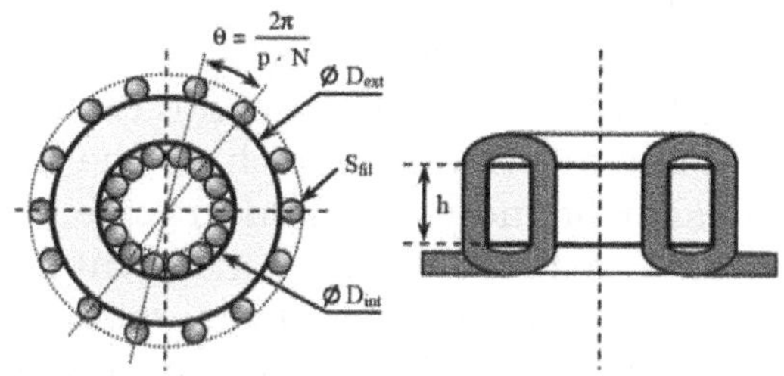

S_{fil}: Section of a conductor
S_b: Internal coiling area
S_{CM}: Cross-section of the magnetic circuit
δ: acceptable current density
N: Number of turns in a winding
p: Number of phases

Figure 4.28. *Distribution of the loops on a toric circuit and inductance data*

The sizing of CMIs is, to a certain degree, similar to that of smoothing inductances. The choice of the magnetic circuit can be carried out using the product (A_i) of the circuit section (S_{CM} or A_e) through the surface of the coiling (S_b). This parameter is defined by the inductance value, the acceptable current density, the saturation induction and the maximum common-mode current [4.16]. This expression leads to an initial observation. The sizing of this component requires the knowledge of the common-mode current generated by the whole of the conversion structure. This is assuming that the filter is equally integrated into the structure. We will then see that the only way to obtain a correctly defined component involves an iterative simulation for estimating the common-mode current at each modification of the components in the filter.

4.4.2.2. *Proposed sizing rule*

The expression of the product of the areas A_i normally leads to a relation that is independent of the number of turns. This parameter is, therefore, not directly managed that can lead to the near-total filling of the winding window. By setting the number of layers, A_i is inevitably a function of *N* through the winding coefficient K_N. A development of this equation, limited over *N* is given by equation [4.17]. The k_f factor defines a security margin, such as the winding factor used in the design of inductors and transformers ($k_f \geq 1$). This approximate expression of K_N is interesting for a number of turns greater than 5, thus raising the relative error to 8% (Figure 4.29). For very high-power converters (e.g. variable-speed drives greater than 30 kW) and particularly in three-phase applications, the CMIs can be created with a very small number of turns for each winding (two to three turns). It would, therefore, also be necessary to consider the exact expression of K_N.

$$A_i = K_N(N) \cdot \frac{L \cdot I_{MC} \cdot I_{eff}}{\delta \cdot B_M} \quad [4.16]$$

$$K_N(N) \approx k_f \cdot \frac{p}{\pi} \cdot \left(\frac{p \cdot N}{\pi} + 2 \right) \quad [4.17]$$

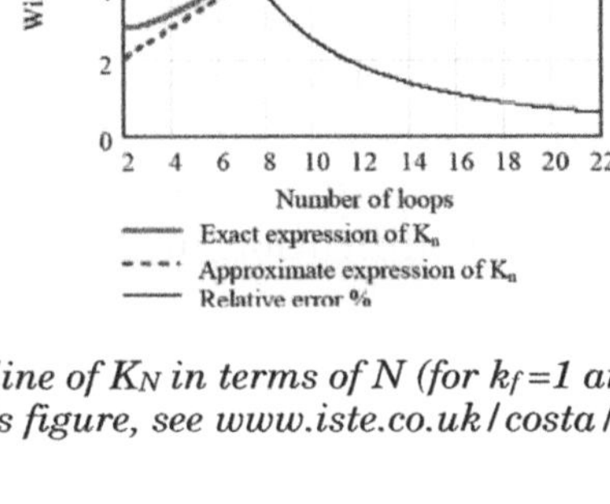

Figure 4.29. *Outline of K_N in terms of N (for k_f=1 and p=2). For a color version of this figure, see www.iste.co.uk/costa/electronics.zip*

We know that the volume of magnetic material can be expressed in terms of the product of the areas product A_i and a dimensionless K_{CM} [4.18]. This factor K_{CM}, often used for the optimization of the volume of magnetic components, depends on the geometric parameters of the circuit (D_{int}, D_{ext} and h). The expression of this term is not unique whether or not we include the winding in volume calculation. By referring to the FerroxCube data for 19 toric circuits from the TX and TN series, the relative permeabilities of which are between 5,000 and 10,500, K_{cm} is defined between 3 and 7. Thus, for the same manufacturer of magnetic circuits, the K_{CM} factor can vary greatly. This shows that the hypothesis for a homothetic sizing of toric inductances, which we based ourselves on, is not always verified.

Nevertheless, in order to simply predetermine the inductance from the FerroxCube data by ensuring a good use of the magnetic circuit, we will set K_{CM} to 3.83, which corresponds to the average value for different cores. From the definition of reluctance, it is possible to establish a relation linking N with the volume of materials (V_{CM}) and with the value of the inductance (L) using the K_V factor (equation [4.20]). This term depends on the permeability of the material and the shape of the magnetic circuit. Using the relations [4.16], [4.17] and [4.20], we can define a polynomial form of A_i (equation [4.21]), the roots of which allow us to determine the product of the areas product A_i, which we are looking for, and to select a circuit from the list of the manufacturer.

$$V_{CM} = K_{CM}.A_i^{3/4} \tag{4.18}$$

$$N = K_v.\sqrt{L}.A_i^{-1/8} \tag{4.19}$$

with $K_v \approx \frac{3.3}{\sqrt{\mu}}$

$$A_i - \alpha . A_i^{-1/8} - \beta = 0 \qquad [4.20]$$

with $\begin{cases} \alpha = \dfrac{4 \cdot k_f}{\pi} \cdot \dfrac{L \cdot I_{MC} \cdot I_{eff}}{\delta \cdot B_M} \\ \beta = \dfrac{4 \cdot k_f \cdot K_V}{\pi^2} \cdot \dfrac{L^{3/2} \cdot I_{MC} \cdot I_{eff}}{\delta \cdot B_M} \end{cases}$

To illustrate this approach, the relative permeability is set to 10,000. This value corresponds to the FerroxCube 3E6 reference. The solutions of equation [4.21] allow us to outline a set of characteristics defining the areas product in terms of the maximum common-mode current and the value of the inductance (Figure 4.30). Thus, for a given material with a known permeability, generally used for high-frequency filtering, the choice of magnetic circuit can be made using a simplified expression [4.22] taken from the network of curves. This expression taken from the linearization of relation [4.21] in terms of L gives very satisfactory results (Figure 4.31).

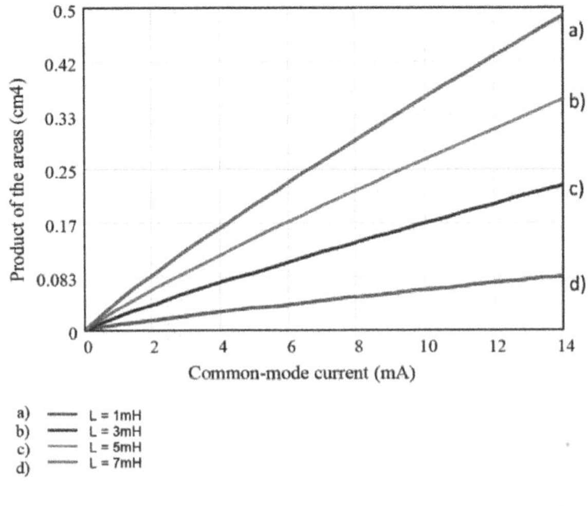

Figure 4.30. *Areas product in terms of IMC and L*

Linearization for L = 3 mH:

$$A_i = C_1(L) \cdot I_{MC} + C_2(L) \quad [4.21]$$

with
$$\begin{cases} C_1(L) = 4.618 \cdot 10^3 \cdot L + 1.829 \\ C_2(L) = 2.952 \cdot L + 1.655 \cdot 10^{-3} \end{cases}$$

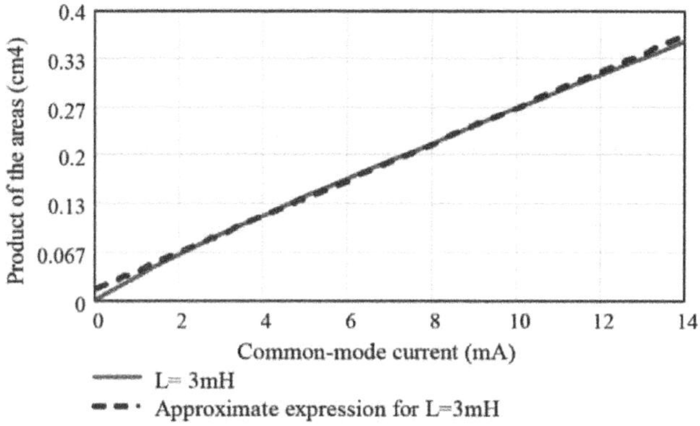

Figure 4.31. *Comparison between the simplified expression and the solution of the polynomial for L = 3mH*

The relation is evidently simpler to use in a process of optimization in which it is preferable to know the direction of variation of A_i in terms of the variables to be optimized such as L. If we want to propose an optimal sizing, it is essential to determine the equivalent impedance and the equivalent common-mode voltage generator of the equipment. Inserting equation [4.22] into an optimization process under certain constraints leads to the choice of the most well-suited magnetic circuit in order to minimize the volume of the inductance by limiting parasitic capacitive effects [SHE 04].

4.4.3. *Results*

The resulting filter, without any real optimization, lets us define a theoretical reference volume. For the inductance, the magnetic circuit has a high permeability and only has one layer of conductors. For the previous spectrum, the required attenuation is of 46 dBμV at 150 kHz (Figure 4.16). The obtained results are recorded in Table 4.2.

	L_{CM}	C_Y
Values	5.1 mH	22 nF
Volume (cm^3)	9.8	2 × 1.9

Table 4.2. *Elements of the reference filter*

Thus, the total volume of the reference common-mode filter is 13.6 cm^3. As an example, an optimization under certain constraints by setting as a goal, the minimization of the volume allows for a relative reduction of a few percent (5–10%) (Table 4.3).

	L_{CM}	C_Y
Values	3.4 mH	33 nF
Volume (cm^3)	7.3	2 × 2.5

Table 4.3. *Elements of the optimized filter*

The objective function, linked to the volume and associated with an optimization algorithm under certain constraints, allows us to create a logical compromise between the volume and the inductance. By setting realistic values to parasitic elements of the components, which we have carefully detailed in section 4.1 , the equivalent model of the calculated filter defines a spectrum, the level of which remains less than the standard profile starting from 150 kHz.

In this study, we have carried out an optimization with a single objective and with the constraint of complying with the standard limitation. This process evidently remains valid for multi-objective optimizations for which we can consider the execution cost or even include thermal aspects by defining the losses associated with each component.

4.5. Conclusion

Unlike the traditional sizing of a differential mode or smoothing inductance, for which the maximum stored energy is known, that of a symmetrical inductance (i.e. in common mode) is more tricky, since it requires knowledge of the maximum amplitude of the common-mode current to avoid saturation of the circuit. Due to this, the knowledge of this current is essential. An *a priori* sizing of this component from a prototype and the usual calculation rules for filters can lead to a malfunction due to saturation of the magnetic circuit. We know that such a process during the development phase can be expensive. Thus, optimization under constraints of the filter, which can also be associated with a global optimization of the structure, can be of vital help in the design of power converters. The sizing technique discussed in this section must allow for the assurance of optimal operation of the filter.

4.5.1. *Corrected exercise: filtering the conducted perturbations of a step-up chopper*

Figure 4.32 shows the spectrum of high-frequency perturbations measured at the input of a sinusoidal BOOST absorber power factor corrector (PFC); it is connected to the 230 V/50 Hz grid. Its electrical and topological diagram is shown in Figure 4.33, with the switching frequency at 85 kHz. The measurement is carried out using the LISN, where we provide the equivalent diagram in Figure 4.36, simplified with regard to the propagation of high-frequency perturbations. We wish to estimate its perturbations regarding the class A EN55011 standard (see graph in Figure 4.32).

We assume that the output voltage is Vs = 400 V and we assume $f_m(t)$ be the modulation function of the chopper.

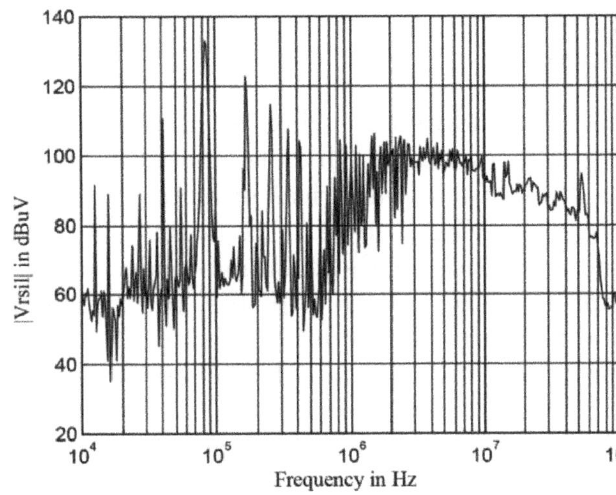

Figure 4.32. *Spectrum of the perturbations measured at the input of the converter at the terminals of the resistance R_{o1} of the LISN*

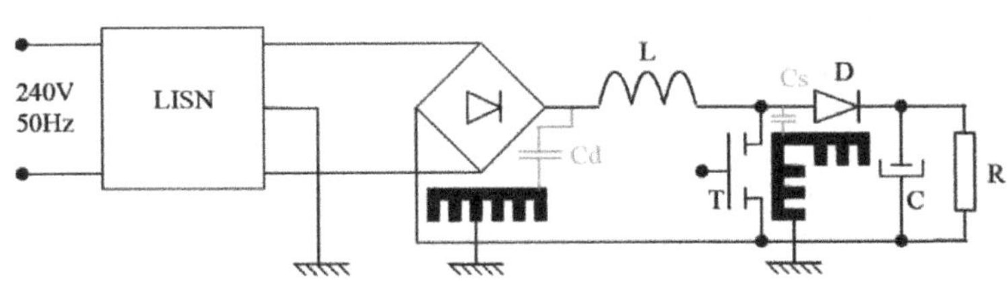

Figure 4.33. *Structure of the converter*

We have characterized the inductance L using a network analyzer up to 30 MHz. The values of the elements are as follows: L = 883 µH; r_p = 19 kΩ; C_p = 105 pF.

1) The resulting spectrum from the parasitic voltage V_{rsil1} measured by the LISN has been outlined in Figure 4.34 (dark gray line). We have also included the spectrum of voltage V_k (α = 0.5, τ = 200 ns) as well as the coupling functions of the class A EN550011 standard.

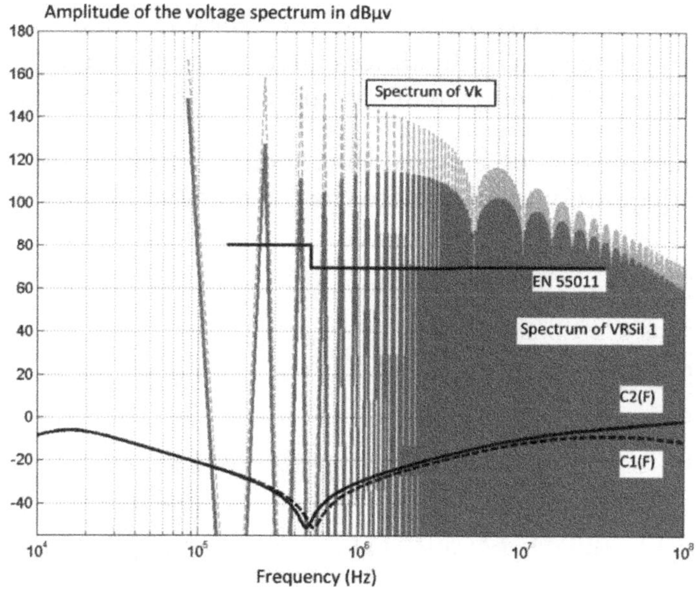

Figure 4.34. *Resulting spectrum of the parasitic voltage V_{rsil1}*

We observe that the spectrum of V_{rsil2} is attenuated in accordance with the coupling function $C_1(F)$, with a low minimum at the resonance frequency of L. In the frequency range from 600 kHz to 4 MHz, the spectrum is at a roughly constant level: this results in the decrease over 1/F of the spectrum of V_k and the increase over F of the function $C_1(F)$ in this range. Beyond this, the decrease over $1/F^2$ of the spectrum of V_k and that of $C_1(F)$ in the range of 10–100 MHz determines the decrease of the spectrum of V_{rsil1}.

The observation of the measured spectrum lets us record its behavior; we have superimposed on Figure 4.35 the actually measured spectrum and the spectrum from the model. The differences are explained by:

– the modulation function of the model, which was assumed constant in the model when in reality it varies with time between 0 and 1;

– the non-consideration of the second-order parasitic effects (connection inductances, output circuit, etc.), which explain the HF evolution of the spectrum (resonance peaks).

Nevertheless, we can see that such a simple model allows us to adequately approximate the phenomena in play.

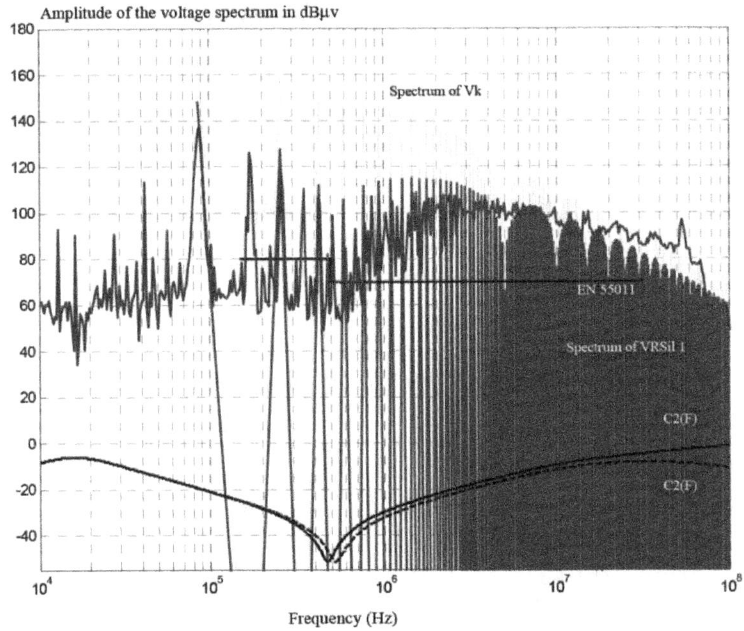

Figure 4.35. *Comparison between the measured spectrum and the spectrum from the model*

2) This perturbation spectrum does not necessarily comply with the EN55011 standard, and the benchmark is exceeded over the whole of the frequency range (150 kHz–30 MHz). It is, therefore, essential to place an EMC filter at the input, the structure of which is shown in Figure 4.36. This must comply with two requirements in accordance with the analysis carried out on the distribution of parasitic currents, and the parasitic components responsible for them:

– Attenuating the differential-mode currents i_{md1} and i_{md2}: this is undertaken by the filter C_x L_x, which is symmetrical over the two power lines.

– Attenuating the common-mode current: this is executed by the capacitors C_y; they constitute a low-value impedance, which allows the current i_{mc} to flow closest to the source V_k.

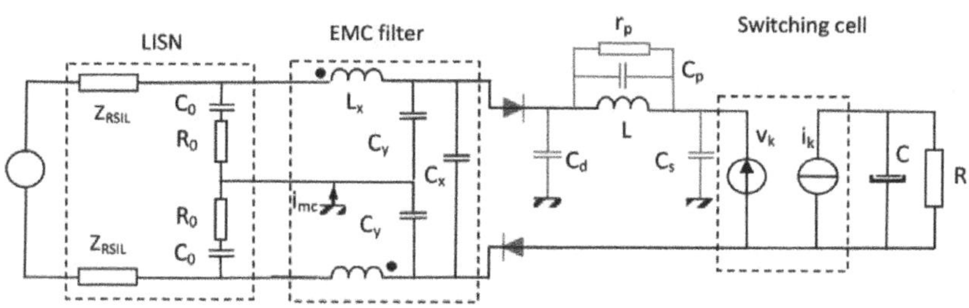

Figure 4.36. *Structure of the EMC filter placed upstream from the converter*

The following is the equivalent simplified diagram:

Remarks on the symmetrical structures of EMC filters

An EMC filter must always have a symmetrical structure in order to limit the transfers between propagation modes. Indeed, we can represent the filter by the diagram below where the impedances of the LISN and those of the filter are assumed to be symmetrical. Thus, the differential-mode voltage V_{diff} creates a zero common-mode current: there is no transfer between the differential and common mode; the filter can be seen as a bridge, the impedances of which are balanced. Similarly, if a common-mode current tends to flow between points O and G (created by other circuits connected elsewhere) and knowing that the ground impedance is never perfectly zero (particularly at HF), it will never produce a differential-mode voltage. This is the reason why we use coupled inductances in order to create the inductances of EMC filters.

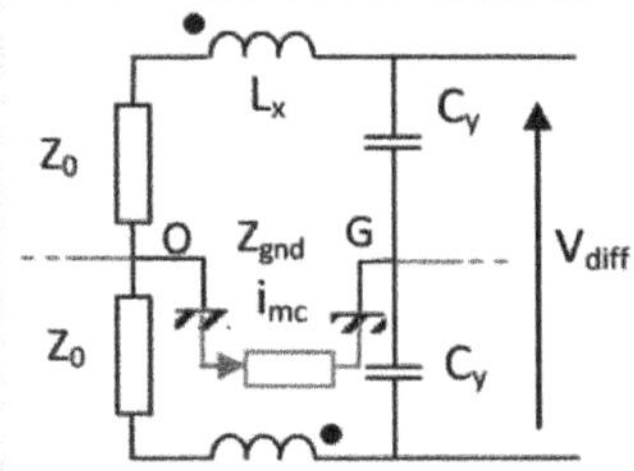

Structure of a symmetrical EMC filter

3) Determining currents in the LISN becomes complicated from an analytical point of view. It is, however, possible to assume that the dominant effects differ according to the frequency range being used.

4.5.1.1. *At high frequency*

We can outline the following equivalent diagram (Figure 4.37), valid at high-frequency (F > 10 MHz) knowing that the impedance C_x is very low in this frequency range, that the impedance of L is supposedly very high and that $Z_0 \neq R_0$:

We deduce that the voltage at the terminals of the impedances of the LISN is expressed as:

$$V_{lisn} = V_k \frac{R_0}{2} \frac{pC_s}{1 + pR_0(C_y + C_s)} \neq V_k \frac{C_s}{2C_y}$$

Thus, if we want a reduction of the order of 20 dB for F > 10 MHz, we choose $C_y \sim 5$ Cs, thus a value greater than 125 pF.

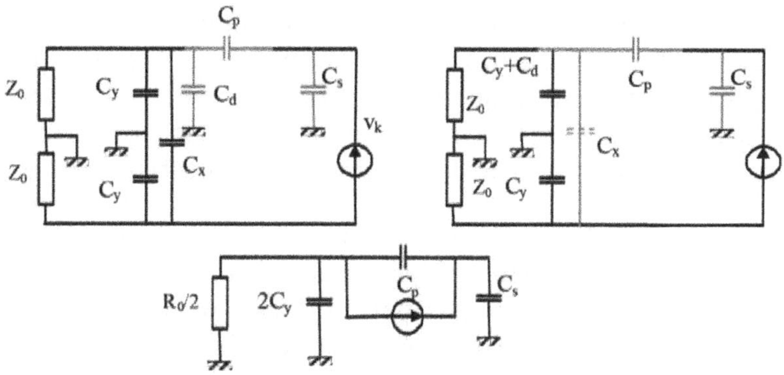

Figure 4.37. *Equivalent diagrams for high frequency*

4.5.1.2. *At low frequency*

For F < 10 MHz, we can see that the two functions $C_1(F)$ and $C_2(F)$ are practically equal within a margin of 2 dB (difference of 25%), which means that the current i_{mc} remains low compared to the others, or even that we can ignore the effect of the capacitances Cd and Cp. We can, therefore, simplify the diagram by taking $i_{md1} = i_{md2}$, which corresponds to the only flow of a differential current in accordance with Figure 4.38.

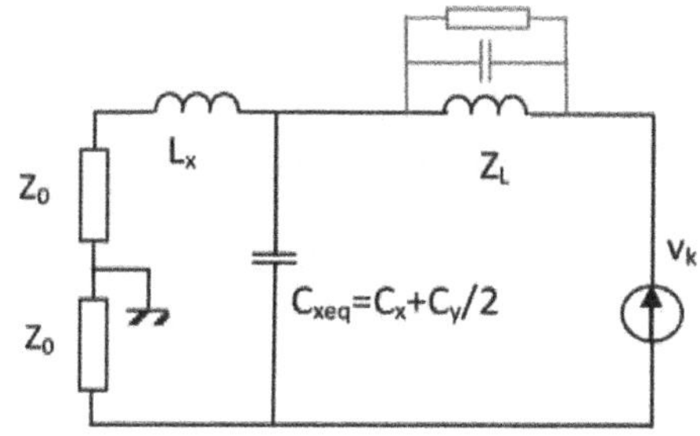

Figure 4.38. *Equivalent diagram for low frequencies*

The equivalent capacitor is equal to $C_x + C_y/2$; effectively, the presence of common-mode capacitors must be taken into account. Therefore, we recalculate the transfer function with the filter, with the values of L_x and C_x being determined to bring a sufficient attenuation starting from a frequency of 150 kHz, which corresponds to the start of the standard limitation. The first harmonic involved is at 255 kHz, and must be attenuated by at least 25 dB (see spectrums).

Nevertheless, we must also take into account the interaction between the components of the filter and those of the converter and their parasitic effects, which will lead to multiple resonances and antiresonances. The return to numeric calculation, even in the simple case presented here, is necessary for adjusting the values of the components of the filter.

Figure 4.39 illustrates these effects. We have chosen L_x = 800 µH, C_x = 47 nF and C_y = 1 nF. We note that the standard profile is complied with; we find the reduction of the parallel resonance frequency of L toward 500 kHz; however, another resonance appears at approximately 900 kHz, which is due to a series resonance between L, L_x and C_p, the parasitic capacitance of L.

A second criterion for the choice of components L_x and C_x consists of researching the minimum value of the filter volume. In general, this is mainly determined by the value of L_x, the process will then consist of finding the minimum value of L_x, then by adjusting the value of C_x in order to ensure the compliance with the standard limitation. Figure 4.40 shows the results obtained with L_x = 10 µH and C_x = 100 nF. The profile complies with standards, the inductance to be placed will be low-volume, its effect can be influenced by the connections from the filter to the LISN, which can be of the same order of magnitude; mastering the behavior of the frequency of the filter is therefore at random, which is why it is preferable to preserve a value that is clearly greater than the eventual connection effects.

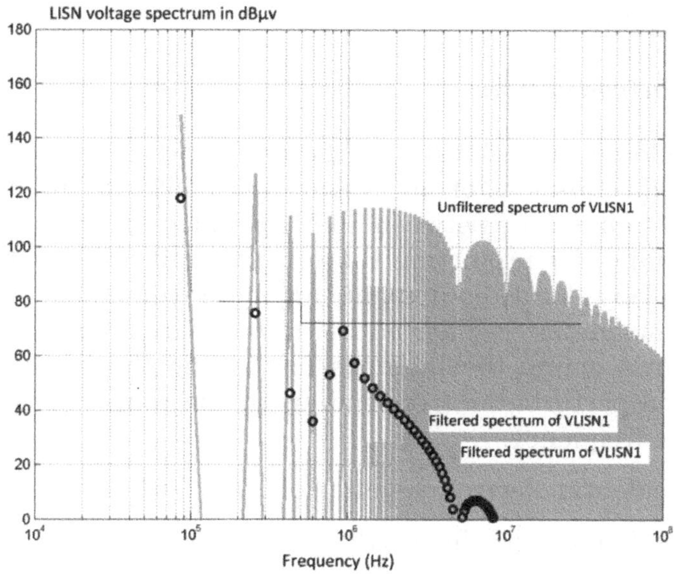

Figure 4.39. *Spectrum of the LISN voltage placed at the input of the converter. Filter 1*

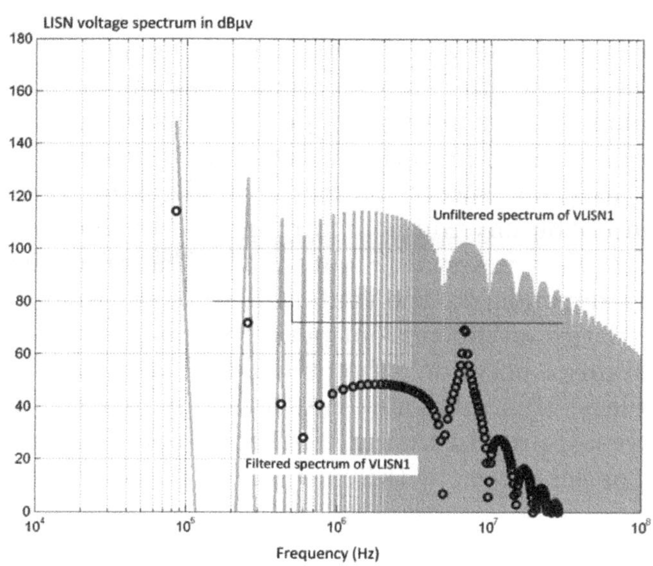

Figure 4.40. *Spectrum of the LISN filter placed at the terminals of the converter. Filter 2*

4.6. Shielding

4.6.1. *Introduction*

Shielding is essential, on the one hand, for limiting the radiated emission of a system (EN55022 standard, for radiated emissions between 30 MHz and 1 GHz), and on the other hand, for protecting the system against perturbations emitted by local equipment or more commonly against electromagnetic radiation in the environment. Several standards define the methods for testing the immunity to radiated perturbations depending on the source of the field and the frequency:

– incidental electromagnetic fields (IEC 61000-4-3);

– 50 Hz magnetic fields (IEC 61000-4-8);

– impulsive magnetic fields (IEC 61000-4-9);

– dampened oscillating magnetic fields (IEC 61000-4-10).

Shielding is generally ensured through a metal or metal-plated casing, which surrounds the equipment. It can also protect a particular area in some equipment such as a switch-mode power supply to the radio section of a DECT telephone. It must protect the sensitive parts of a material against exterior electromagnetic fields and limit their pollution of the environment through electromagnetic fields created by this material (Figure 4.41). The efficiency of the shielding depends on the material used, its thickness, the characteristics of the current perturbation source as well as the distance between the shielding and this source. Nonetheless, this efficiency can be hampered by openings, which are required to ensure ventilation and the expulsion of heat in the casing, or by the connections of power or interface cables.

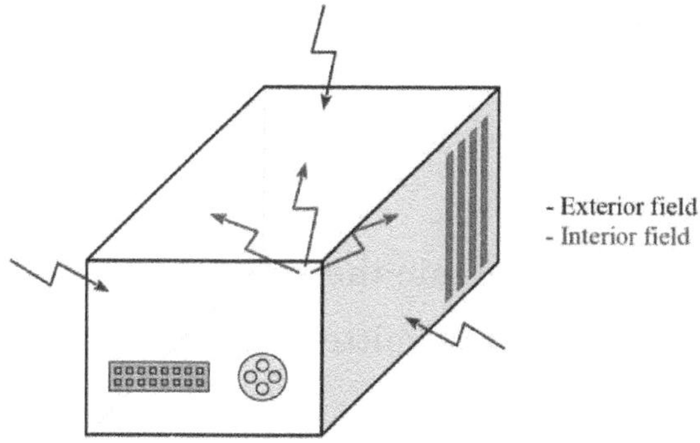

Figure 4.41. *Metallic casing used for shielding equipment. For a color version of this figure, see www.iste.co.uk/costa/electronics.zip*

From a physics point of view, the two effects that we will present are both reflections, which are produced when a wave passes from the air through a conductive space or inversely, and attenuation of a wave when it is propagated in a conductive space.

4.6.2. *Breakdown of shielding effects*

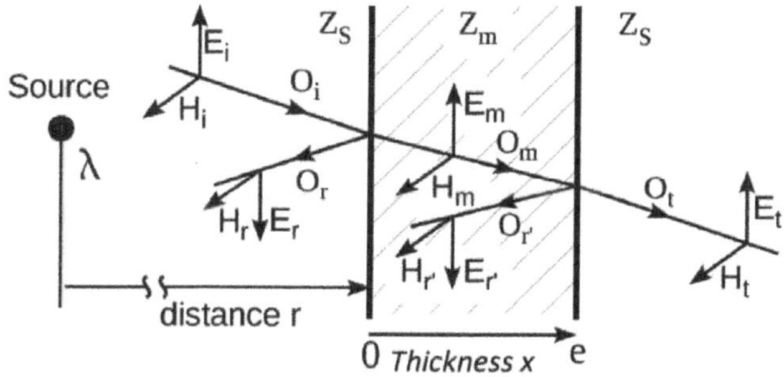

Figure 4.42. *Flow of an electromagnetic wave through shielding*

The distance from the source is noted as r and its wavelength λ. A reflection is produced when a wave arrives at the shielding (x = 0). Then, the wave O_m is propagated in the shielding and is attenuated. Lastly, a second reflection is produced when the wave leaves the shielding to set off into the air (x = e).

– Air/metal interface, x = 0:

- O_i: incidental wave;

- O_r: wave reflected into the air;

- $O_m(0)$: wave transmitted upon entrance to the shielding.

– Position x:

- $O_m(x)$: propagation of the wave throughout the material, attenuation.

– Air/metal interface, x = e:

- $O_m(e)$: incidental wave leaving the shielding;

- $O_{r'}$: wave reflected in the material;

- O_t: wave transmitted upon leaving the shielding.

These effects mainly rely on the characteristics of the material used for the shielding and the nature of the electromagnetic field upon entering the shielding (near field, E or H dominant, or far field, electromagnetic wave), which are characterized by:

– Z_s: impedance of the wave in the air (at a distance r from the source);

– Z_m: impedance of the wave in the material (depends on ω, μ and σ);

– δ: skin depth (depends on ω, μ and σ).

4.6.2.1. *Attenuation linked to the reflections R_{dB}*

At the interface between the air and the material of the shielding, the reflective effects are linked to the mismatch between the wave impedance Z_s at the interface level and the material's impedance Z_m. The term R_{dB} encompasses the two reflective effects upon entrance and exit of the shielding without taking into account the propagation of the wave in the shielding, which could correspond to the case of an extremely thin shielding.

4.6.2.2. *Attenuation linked to the absorption A_{dB}*

When the wave travels through the material, its amplitude decreases exponentially. This attenuation is linked to the skin depth and the thickness of the coating at the frequency being used.

The total attenuation S_E is the sum of these two effects:

$$S_{EdB} = A_{dB} + R_{dB} \tag{4.22}$$

When using materials with a low conductivity or multilayer shielding, it is possible that an additional attenuation B_{dB} must be considered, linked to the multiple reflections within the shielding; however, we will not elaborate on this in this section. Figure 4.43 shows the global attenuation for a copper shielding plane with a thickness of 2 mm in far field.

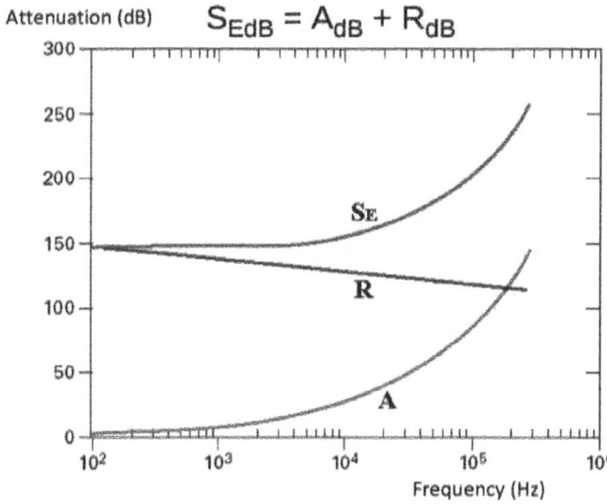

Figure 4.43. *Global attenuation for 2 mm copper shielding in far field*

4.6.3. *Materials*

If we consider a homogenous conductive material, its most important characteristic is its conductivity σ [$S \cdot m^{-1}$]. If the material has magnetic properties, its magnetic permeability μ [$H \cdot m^{-1}$] will be its second most important characteristic. In the following, we will only consider the shielding in the case of homogenous materials that have a linear behavior for the commonly used electromagnetic field amplitudes (Table 4.5). The relative permeability μ_r is defined relative to the vacuum permeability μ_0, and the relative conductivity G is defined in relation to the conductivity of the copper.

Constants and physical properties			
μ_0	=	$4\pi \times 10^{-7}$	$[\mathrm{H} \cdot \mathrm{m}^{-1}]$
ε_0	=	$1/(36\pi) \times 10^{-9}$	$[\mathrm{F} \cdot \mathrm{m}^{-1}]$
c	=	3×10^8	$[\mathrm{m} \cdot \mathrm{s}^{-1}]$
σ_{copper}	=	5.8×10^7	$[\mathrm{S} \cdot \mathrm{m}^{-1}]$

Table 4.4. *Recall of the constants and physical properties*

$$\mu = \mu_0 \times \mu_r \qquad [4.23]$$

$$\sigma = G \times \sigma_{copper} \qquad [4.24]$$

4.6.3.1. *Physical characteristics*

Among non-magnetic materials, silver has the best conductivity, however it is not used for obvious cost reasons. Copper serves as the reference material, and it is sometimes used to create outer casings, with problems linked to oxidation. Nonetheless, the most adopted non-magnetic material used in the industry is aluminum for its cost/conductivity ratio. Among the materials having magnetic properties, steel is used frequently, particularly in power electronics. Because of its magnetic permeability, it can sometimes be more efficient at the same thickness as copper in low frequency, under certain conditions, which we will specify later [HEI 80].

Materials	Relative conductivity in relation to copper	Relative permeability (100 Hz)
	G	µr
Silver	1.05	1
Annealing copper	1.00	1
Gold	0.70	1
Aluminum	0.61	1
Brass	0.26	1
Nickel	0.20	1
Bronze	0.18	1
Tin	0.15	1
Steel, 3% Si	0.10	1,000
Lead	0.08	1
Monel	0.04	1
Stainless steel	0.02	500
78 Permalloy	0.11	15,000
Mu-metal	0.03	13,000

Table 4.5. *Relative conductivity and magnetic permeability for different materials*

However, the relative permeability of steel decreases with the frequency as shown in Table 4.6, which reduces its efficiency for frequencies more than 100 MHz.

Steel, 3% Si									
Frequency	kHz	MHz					GHz		
	≤10	1	3	10	30	100	1	1.5	10
μr	1,000	700	550	400	300	200	50	10	1

Table 4.6. *Magnetic permeability of steel in terms of the frequency*

Certain materials such as Permalloy 78 or Mu-metal are particularly well adapted to shielding in the case of magnetic fields in low frequency. Nonetheless, for these materials, the relative magnetic permeability, which is very high for frequencies less than 100 Hz, decreases quickly with the frequency as shown in Figure 4.44.

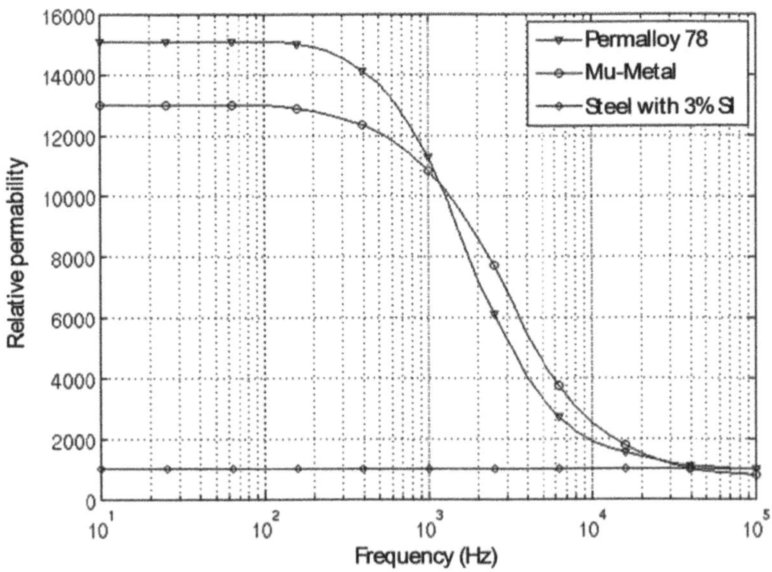

Figure 4.44. *Evolution of the relative permeability as a function of the frequency*

4.6.3.2. *Wave impedance*

The relation between the transverse electrical field and the transverse magnetic field in the case of a wave that is propagated in a conductive material is determined by the properties of the material. In most cases, the wave impedance in the material Z_m, defined by equation [4.25], depends also on the dielectric permittivity ε (equation [4.26]).

$$Z_m = \frac{|\vec{E_t}|}{|\vec{H_t}|} \qquad [4.25]$$

$$Z_m = \sqrt{\frac{j\omega\mu}{\sigma + j\omega\varepsilon}} \qquad [4.26]$$

However, in the case of conductive materials, the term $\omega\varepsilon$ remains extremely low compared to the conductivity σ even for frequencies more than several gigahertz, which allows for the simplification of the expression of the modulus of Z_m (equation [4.27]).

$$|Z_m| = \sqrt{\frac{\omega\mu}{\sigma}} \qquad [4.27]$$

4.6.3.3. *Skin depth*

The skin depth δ allows us to determine from which pulsation ω the distribution of the current in a massive conductive material of a given diameter or thickness is no longer homogenous. If, for a given pulsation ω, the skin depth is lower than the size of the conductor, then the current will be spread across the edge of the conductor. For conductive materials, the skin depth is defined by the equation:

$$\delta = \sqrt{\frac{2}{\omega\mu\sigma}} \qquad [4.28]$$

A simple example is that of the distribution of current in a cylindrical conductor, such as the one shown in Figure 4.45. At a frequency of 10 Hz, the skin depth for aluminum is 27 mm. The top section represents the distribution of the current in the conductor, with the darker gray color corresponding to the highest density of the current, and the lower section represents the density module of the current (in A/mm²) following a line passing through the center. We can see in the figure that the majority of the current is distributed around the outer edge of the conductor.

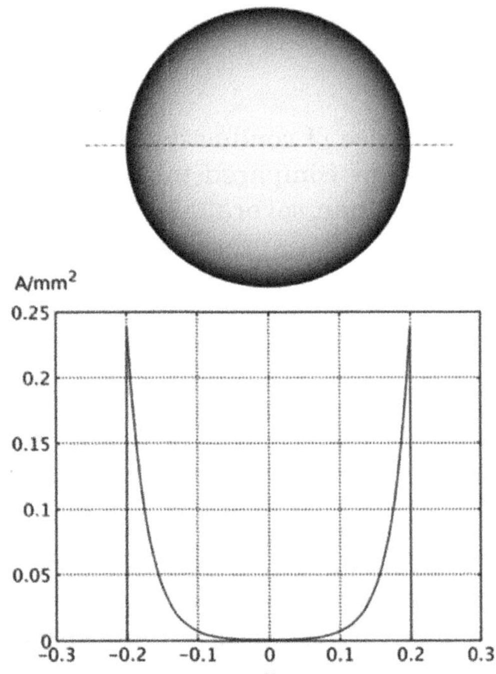

Figure 4.45. *Skin depth at 10 Hz for an aluminum conductor*

When the wave penetrates the metal, the ratio between the shielding thickness and skin depth at the considered frequency will define the undergone attenuation.

4.6.4. *Wave impedance*

4.6.4.1. *High-impedance source – electrical field*

We consider the electrical field created by the oscillating dipole with dipolar moment $\vec{p}$ and pulsation $\omega = 2\pi f$. We associate this pulsation with the wavelength $\lambda = c/f$. The distance from the source is noted as r, and the local vector field in which the electrical and magnetic fields are expressed is shown in Figure 4.46. The exact equations of the fields are given below (equations [4.30] and [4.31]).

$$\vec{p} = q.\overrightarrow{AB} \qquad [4.29]$$

$$\overrightarrow{E_r} = \frac{p.\cos(\theta)}{2\pi\varepsilon_0 r^3}\left(1 + j.\frac{2\pi r}{\lambda}\right)e^{-j.\frac{2\pi r}{\lambda}}.\overrightarrow{u_r}$$

$$\overrightarrow{E_\theta} = \frac{p.\sin(\theta)}{4\pi\varepsilon_0 r^3}\left(1 + j.\frac{2\pi r}{\lambda} + j^2.\frac{4\pi^2 r^2}{\lambda^2}\right)e^{-j.\frac{2\pi r}{\lambda}}.\overrightarrow{u_\theta} \qquad [4.30]$$

$$\overrightarrow{E_\phi} = 0.\overrightarrow{u_\phi}$$

$$\overrightarrow{H_r} = 0.\overrightarrow{u_r}$$

$$\overrightarrow{H_\theta} = 0.\overrightarrow{u_\theta} \qquad [4.31]$$

$$\overrightarrow{H_\phi} = \frac{j\omega.p.\sin(\theta)}{4\pi r^2}\left(1 + j.\frac{2\pi r}{\lambda}\right)e^{-j.\frac{2\pi r}{\lambda}}.\overrightarrow{u_\phi}$$

For the dipole oscillating with the orientation defined in Figure 4.46, the electric field has no component in direction $\overrightarrow{u_\phi}$ and the magnetic field has no component in directions $\overrightarrow{u_r}$ and $\overrightarrow{u_\theta}$.

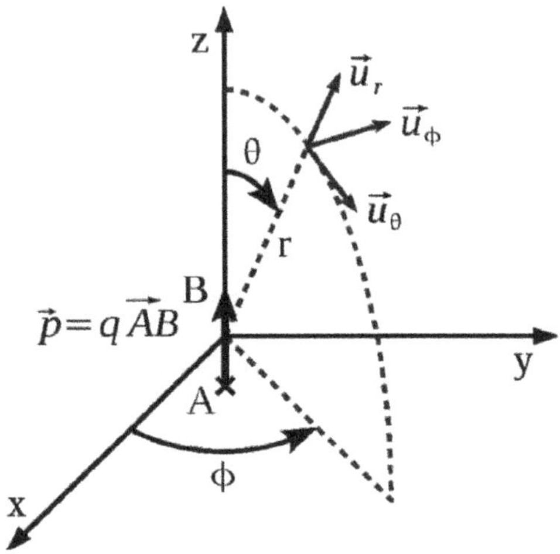

Figure 4.46. *Oscillating dipole: local field for the expression of the electrical field*

4.6.4.2. *Low-impedance source – magnetic field*

Similarly, we can express the magnetic field created by a radiating loop with magnetic moment $\vec{M}$, still in the local vector field, shown in Figure 4.47. In this configuration, the equations of fields are known and recalled here (equations [4.33] and [4.34]).

$$\vec{M} = I.S.\vec{n} \quad [4.32]$$

$$\overrightarrow{H_r} = \frac{IS.\cos(\theta)}{2\pi r^3}\left(1 + j.\frac{2\pi r}{\lambda}\right)e^{-j.\frac{2\pi r}{\lambda}}\,\overrightarrow{u_r}$$

$$\overrightarrow{H_\theta} = \frac{IS.\sin(\theta)}{4\pi r^3}\left(1 + j.\frac{2\pi r}{\lambda} + j^2.\frac{4\pi^2 r^2}{\lambda^2}\right)e^{-j.\frac{2\pi r}{\lambda}}\,\overrightarrow{u_\theta} \quad [4.33]$$

$$\overrightarrow{H_\phi} = 0.\overrightarrow{u_\phi}$$

$$\overrightarrow{E_r} = 0.\overrightarrow{u_r}$$

$$\overrightarrow{E_\theta} = 0.\overrightarrow{u_\theta} \qquad [4.34]$$

$$\overrightarrow{E_\phi} = \frac{j\omega.\mu_0.IS.\sin(\theta)}{4\pi r^2}\left(1 + j.\frac{2\pi r}{\lambda}\right)e^{-j.\frac{2\pi r}{\lambda}}.\overrightarrow{u_\phi}$$

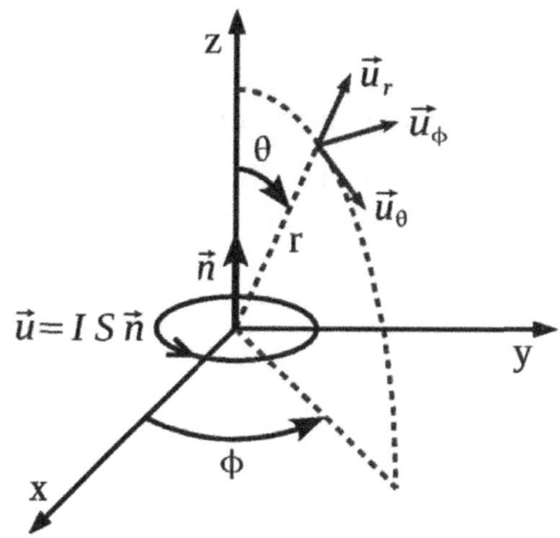

Figure 4.47. *Radiating loop: local field for the expression of the magnetic field*

Similarly, with this orientation of the radiating loop, the magnetic field has no vectorial component along $\overrightarrow{u_\phi}$, and the electric field has no component along $\overrightarrow{u_r}$ and $\overrightarrow{u_\theta}$.

4.6.4.3. *Near-field / far-field limit*

These two theoretical cases let us determine the limit between the near-field and far-field r_{limit} commonly used in shielding calculations.

$$r_{limit} = \frac{\lambda}{2\pi} \qquad [4.35]$$

Nevertheless, this expression is not valid in most cases, and the real limit between the near field and far field varies depending on the source or the antenna being used. However, we will use this expression further on to set a theoretical limit. We define the coefficient k (equation [4.36]) to obtain:

$$k = \frac{2\pi}{\lambda} \quad [4.36]$$

– for $k.r \ll 1$, we are in the near-field zone, with a dominant field depending on the nature of the source;

– for $k.r \gg 1$, we are in the far-field zone, with an electromagnetic wave being formed and a wave impedance of 120π in the air.

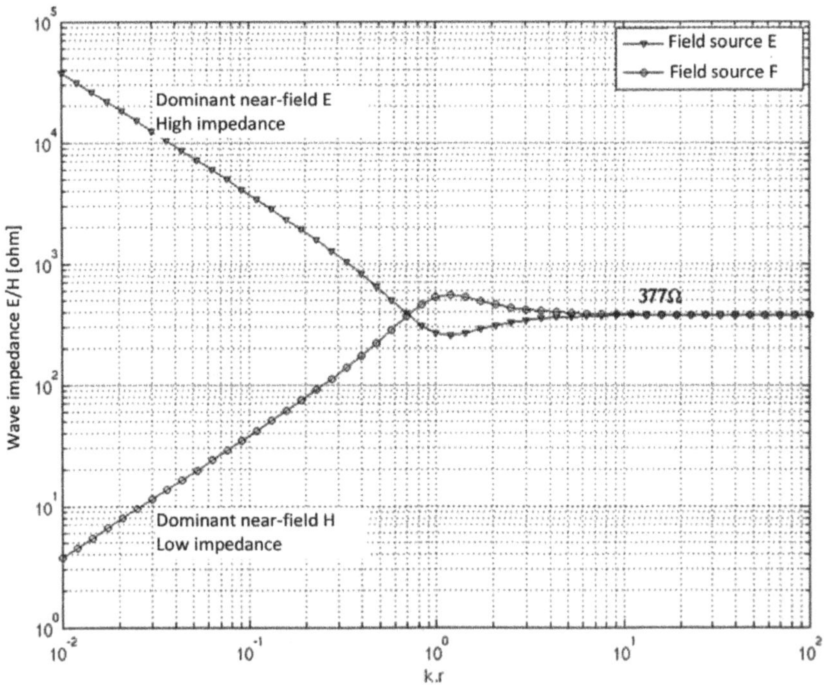

Figure 4.48. *Limit between near field and far field*

4.6.4.4. *Wave impedance in far field*

For $k \cdot r \gg 1$, we can simplify the expressions of the components' electrical and magnetic field modules in order to find the impedance of an electromagnetic wave in the vacuum in far field. This wave impedance is given by the ratio of transverse electrical and magnetic fields. In the case of an oscillating dipole, by simplifying the terms in r:

$$|Z_s| = \frac{|\overrightarrow{E_\theta}|}{|\overrightarrow{H_\phi}|} \quad [4.37]$$

$$|E_\theta| = \frac{p.sin\,(\theta)}{4\pi\varepsilon_0 r} \cdot \frac{4\pi^2}{\lambda^2} \quad [4.38]$$

$$|H_\phi| = \frac{\omega.p.sin\,(\theta)}{4\pi r} \cdot \frac{2\pi}{\lambda} \quad [4.39]$$

Replacing ω with c·2π/λ:

$$|H_\phi| = \frac{c.p.sin\,(\theta)}{4\pi r} \cdot \frac{4\pi^2}{\lambda^2} \quad [4.40]$$

$$|Z_s| = \frac{1}{c.\varepsilon_0} = 120\pi \quad [4.41]$$

In the case of a radiating loop, by simplifying the terms in *r*:

$$|Z_s| = \frac{|\overrightarrow{E_\phi}|}{|\overrightarrow{H_\theta}|} \quad [4.42]$$

$$|E_\phi| = \frac{\omega.\mu_0.I.S.sin\,(\theta)}{4\pi r} \cdot \frac{2\pi}{\lambda} \quad [4.43]$$

$$|H_\theta| = \frac{I.S.sin\,(\theta)}{4\pi r} \cdot \frac{4\pi^2}{\lambda^2} \quad [4.44]$$

Similarly to before, by replacing ω with c·2π/λ:

$$|E_\phi| = \frac{c.\mu_0.I.S.sin\,(\theta)}{4\pi r} \cdot \frac{4\pi^2}{\lambda^2} \quad [4.45]$$

$$|Z_s| = \mu_0.c = 120\pi \quad [4.46]$$

These results let us recall that in far field, in the vacuum, the fields $\vec{E}$ and $\vec{H}$ are combined to form an electromagnetic wave of impedance $Z_0 = 377\,\Omega$, and that the amplitude of these fields decreases over $1/r$.

4.6.4.5. *Electrical near field – high impedance*

For $k \cdot r \ll 1$, it is also possible to simplify the expression of the components' electrical and magnetic field modules. By only taking the field components that are not null, this gives:

$$|E_r| = \frac{p.\cos(\theta)}{2\pi\varepsilon_0 r^3} \quad [4.47]$$

$$|E_\theta| = \frac{1}{\varepsilon_0 r}.\frac{p.sin(\theta)}{4\pi r^2} \quad [4.48]$$

$$|H_\phi| = \frac{2\pi c}{\lambda}.\frac{p.sin(\theta)}{4\pi r^2} \quad [4.49]$$

$$|Z_s| = \frac{|\vec{E_\theta}|}{|\vec{H_\phi}|} = \frac{\lambda}{2\pi.r.c.\varepsilon_0} \quad [4.50]$$

By using the term $k.r$ in the expression Z_s:

$$|Z_s| = \frac{1}{k.r} \times \frac{1}{c.\varepsilon_0} = \frac{1}{k.r} \times Z_0 \quad [4.51]$$

In near field, the dominant electrical field components decrease over $1/r^3$ whereas the magnetic field components decrease over $1/r^2$. As a result, the wave impedance is increased close to the source and evolves over $1/r$ once it leaves.

4.6.4.6. *Magnetic near field – low impedance*

In the case of a radiating loop, invariably for $k.r \ll 1$, the simplified expressions of the components' electrical and magnetic field modules are:

$$|H_r| = \frac{I.S.\cos(\theta)}{2\pi r^3} \quad [4.52]$$

$$|H_\theta| = \frac{1}{r}.\frac{I.S.sin\,(\theta)}{4\pi r^2} \qquad [4.53]$$

$$|E_\phi| = \frac{2\pi c.\mu_0}{\lambda}.\frac{I.S.sin\,(\theta)}{4\pi r^2} \qquad [4.54]$$

$$|Z_s| = \frac{|\overrightarrow{E_\phi}|}{|\overrightarrow{H_\theta}|} = \frac{2\pi.r.\mu_0.c}{\lambda} \qquad [4.55]$$

In the same way as the oscillating dipole, we can express the wave impedance in terms of $k.r$ and Z_0 (equation [4.56]).

$$|Z_m| = k.r \times Z_0 \qquad [4.56]$$

It is the dominating magnetic field components that decrease over $1/r^3$, whereas the electrical field components decrease over $1/r^2$. As a result, the wave impedance is low when close to the source, and it increases over r as it furthers itself from it.

4.6.4.7. *Review of the wave impedance*

The previous results are summarized in Table 4.7 for a wave of given pulsation ω in terms of the product k·r, with k = 2π/λ and r the distance between the source and the observation point.

Wave	Wave impedance	
Far field	$Z_s = 120\pi$	$k.r \gg 1$
Near field, E source	$Z_s = \frac{1}{k.r} \times 120\pi$	$k.r \ll 1$
Near field, H source	$Z_s = k.r \times 120\pi$	$k.r \ll 1$

Table 4.7. *Wave impedance in terms of the nature of the electromagnetic field*

These three configurations lead to attenuations through different reflections for similar shielding characteristics. Furthermore, if we focus our attention to the efficiency of the shielding over a large frequency range, in the case of a source that is relatively close to the shielding, it is possible to end up in near-field conditions for lower frequencies and to transfer to far-field conditions when $f \gg c/2\pi r$. For example, for a source that is 1 m from the shielding, the transition frequency is 47.7 MHz.

4.6.5. *Expression of attenuations*

By taking the configuration shown in Figure 4.49, and considering the observations on the materials and waves presented in the previous sections, we will be able to calculate expressions for the attenuations A_{dB} and R_{dB}.

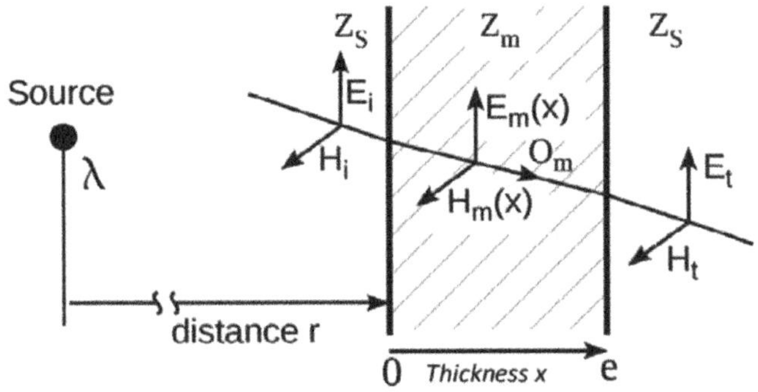

Figure 4.49. *Global configuration of a wave passing through shielding*

4.6.5.1. *Attenuation caused by reflections: R_{dB}*

The attenuation caused by the reflections is divided in two parts: on the one hand, by the reflection of the wave on the air/metal interface upon entrance to the shielding, which will be marked with a high index 0 to remind us that it is located at position x = 0, and on the other hand, by the reflection on the air/metal interface upon leaving the shielding, for x = e, marked with a high index e.

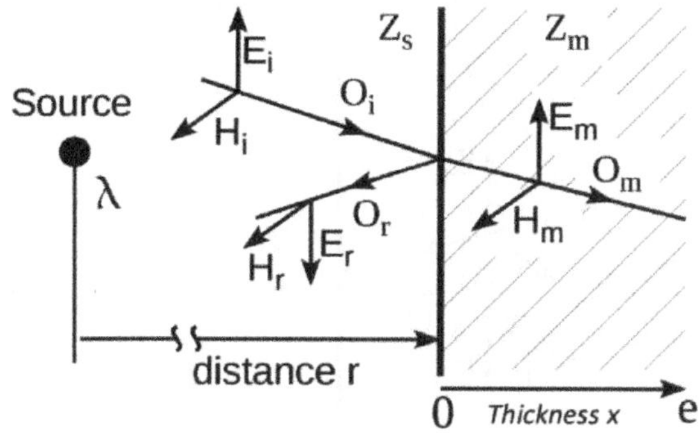

Figure 4.50. *Reflection at the air / metal interface, x = 0*

The reflection coefficient at x = 0 is written as ρ_{AM}, and its expression is given by equation [4.57].

$$\rho_{AM} = \frac{Z_m - Z_s}{Z_m + Z_s} \quad [4.57]$$

Using this refection coefficient, the amplitudes of the reflected and transmitted fields can be expressed as:

$$E_r^0 = \rho_{AM} . E_i^0 \quad [4.58]$$

$$H_r^0 = -\rho_{AM} . H_i^0 \quad [4.59]$$

$$E_m^0 = E_i^0 + E_r^0 = (1 + \rho_{AM}) . E_i^0 \quad [4.60]$$

$$E_m^0 = \frac{2Z_m}{Z_m + Z_s} . E_i^0 \quad [4.61]$$

$$H_m^0 = (1 - \rho_{AM}) . H_i^0 \quad [4.62]$$

$$H_m^0 = \frac{2Z_s}{Z_m + Z_s} . H_i^0 \quad [4.63]$$

We must note that the electrical field and the magnetic field do not undergo the same attenuation between the incidental field and the transmitted field during this reflection, which is consistent since the E/H ratio must change between the air and the metal.

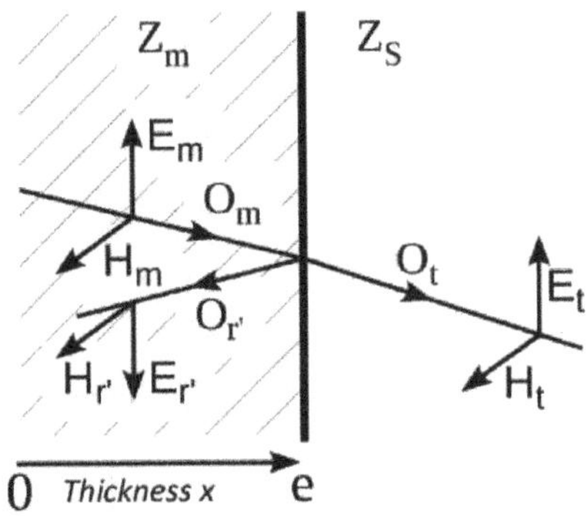

Figure 4.51. *Reflection at the air / metal interface, x = e*

The reflection coefficient ρ_{MA} for the air/metal interface is given by equation [4.64]. The electrical field E_m^e is associated with the wave propagated in the shielding and arrives at the exit interface of the shielding at x = e.

$$\rho_{MA} = \frac{Z_s - Z_m}{Z_s + Z_m} \quad [4.64]$$

$$E_t^e = \frac{2Z_s}{Z_m + Z_s} \cdot E_m^e \quad [4.65]$$

$$H_t^e = \frac{2Z_m}{Z_m + Z_s} \cdot H_m^e \quad [4.66]$$

The attenuations undergone by the electrical and magnetic fields are reversed compared to those calculated for the air/metal interface. If we combine the effects of the two attenuations, the result is then the same for both the magnetic and electrical fields:

$$\frac{E_m^0}{E_i^0} \times \frac{E_t^e}{E_m^e} = \frac{2Z_m}{Z_m + Z_s} \cdot \frac{2Z_s}{Z_m + Z_s} = \frac{4.Z_m.Z_s}{(Z_m + Z_s)^2} \quad [4.67]$$

$$\frac{H_m^0}{H_i^0} \times \frac{H_t^e}{H_m^e} = \frac{2Z_s}{Z_m+Z_s} \cdot \frac{2Z_m}{Z_m+Z_s} = \frac{4.Z_s.Z_m}{(Z_m+Z_s)^2} \qquad [4.68]$$

The two fields are attenuated in an identical manner due to the reflections while passing through the shielding. This result defines the global attenuation R, given by equation [4.69], and its expression in decibels, R_{dB}, equation [4.70].

$$R = \frac{4.Z_m.Z_s}{(Z_m+Z_s)^2} \qquad [4.69]$$

$$R_{dB} = -20.\log_{10}\left(\frac{4.Z_m.Z_s}{(Z_m+Z_s)^2}\right) \qquad [4.70]$$

In the scenario where the impedance of the material is low compared to the wave impedance, this expression can be approximated by equation [4.71] (however, it is crucial that this condition be verified before using this approximation).

$$Z_m \ll Z_s \Rightarrow R_{dB} \approx -20.\log_{10}\left(\frac{4.Z_m}{Z_s}\right) \qquad [4.71]$$

This approximation is often valid in near field for a dominant electrical field or in far field, or even when the frequency is beyond megahertz. It must not be used in the case of a near-field low-frequency magnetic field. The purpose of this approximation is to be able to write simplified expressions for the attenuation R_{dB} in each of these three cases:

Near field H $\quad R_{dB} \approx 14{,}6 - 10.\log_{10}\left(\frac{\mu_r}{fr^2G}\right)$ [4.72]

Near field E $\quad R_{dB} \approx 321{,}7 - 10.\log_{10}\left(\frac{\mu_r f^3 r^2}{G}\right)$ [4.73]

Far field $\quad R_{dB} \approx 168{,}1 - 10.\log_{10}\left(\frac{\mu_r f}{G}\right)$ [4.74]

4.6.5.2. *Attenuation through absorption: A_{dB}*

In the shielding, the magnitude of both electrical and magnetic fields of the wave are attenuated exponentially in terms of the position x, defined using the air/metal interface in the shielding:

$$E_m(x) = E_m(0).e^{-\frac{x}{\delta}} \quad [4.75]$$

$$H_m(x) = H_m(0).e^{-\frac{x}{\delta}} \quad [4.76]$$

with δ the skin depth defined in section 4.6.3.3, whose expression we recall here:

$$\delta = \sqrt{\frac{2}{\omega\mu\sigma}} \quad [4.77]$$

For a shielding thickness e, the attenuation caused by the absorption expressed in decibels is given by equation [4.79].

$$A_{dB} = -20.log_{10}\left(\frac{E_m(e)}{E_m(0)}\right) \quad [4.78]$$

$$A_{dB} = -20.log_{10}\left(e^{-\frac{e}{\delta}}\right) \quad [4.79]$$

This expression can be simplified (equation [4.80]) or be given in terms of the thickness of the shielding "e" and of μ_r, f and G (equation [4.81]).

$$A_{dB} = \frac{-20}{\ln(10)} \times \frac{-e}{\delta} = 8{,}686 \times \frac{e}{\delta} \quad [4.80]$$

$$A_{dB} = 131{,}5.e.\sqrt{\mu_r f G} \quad [4.81]$$

If we set a given thickness "e", we usually estimate that the frequency, beyond which the attenuation through absorption will start to be important, is that for which the skin depth δ is equal to "e".

4.6.5.3. *Global attenuation*

The two forms of A_{dB} and R_{dB} to remember are:

$$A_{dB} = -20.\,log_{10}\left(e^{-\frac{e}{\delta}}\right)$$

$$R_{dB} = -20.\,log_{10}\left(\frac{4.\,Z_m.\,Z_s}{(Z_m + Z_s)^2}\right)$$

Indeed, with the means of calculation available today, it is better to stick to the exact and significant forms of these equations rather than simplified ones. Once the formulas are established, we must define the problem that we wish to study. If, for example, we take interest in the immunity of a system with regard to radiated fields, we would rather adopt far-field conditions, which correspond to distant sources. However, if we wish to limit the radiated field within a switch-mode power supply, the electrical sources of the magnetic field linked to the operation of the power supply will be close to the shielding and we will need to adopt near-field conditions. Finally, these equations exclusively outline the attenuation caused by a perfectly sealed casing without leaks linked to openings or cables. Therefore, they are maximal attenuations that can be very high theoretically, but can decrease quickly. In the following presentation of results, we will limit ourselves to a maximum attenuation of 400 dB.

4.6.6. *Global attenuation: case study*

We will study the efficiency of shielding for aluminum and for steel, with the same shielding thickness, in different conditions. Aluminum has characteristics similar to copper (very good conductivity, and relative permeability $\mu_r = 1$) and a similar behavior. Steel is the reference magnetic material for shielding, with a high relative permeability, which decrease with the frequency, and a very good conductivity, although 10 times lower than that of copper. Finally, the thickness of the shielding will have a strong influence on the effect of attenuation through absorption.

4.6.6.1. *Far-field attenuation for aluminum and steel*

Figure 4.52 shows the global attenuation S_{dB} for aluminum and steel, on the one hand with a thickness of 10 µm (for example in the case of a metallized plastic) and on the other hand for a thickness of 1 mm. The skin depth and the wave impedance Z_m of at 10 kHz are given in Table 4.8. With the wave impedance of aluminum being much lower than that of steel, the attenuation caused by the reflections at 10 kHz is greater in the case of aluminum (126 dB) than it is in the case of steel (88 dB).

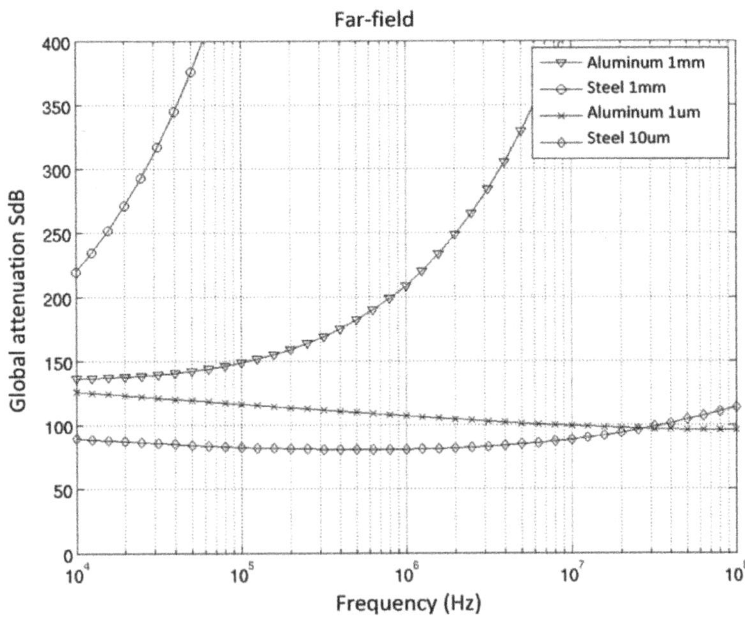

Figure 4.52. *Far-field attenuation for aluminum and steel*

For a thickness of 10 µm, the losses through absorption remain low over the whole frequency range, whereas for a thickness of 1 mm, they dominate in the case of steel and they increase quickly in the case of aluminum. Indeed, at 10 kHz, the skin depth of steel is already much lower than 1 mm, whereas that of aluminum is of the same order of magnitude (Table 4.8).

	Skin depth (10 kHz)	Wave impedance Z_m (10 kHz)	R_{dB}
Aluminum	850×10^{-6} m	$47 \times 10^{-6}\ \Omega$	126 dB
Steel	66×10^{-6} m	$3.7 \times 10^{-3}\ \Omega$	88 dB

Table 4.8. *Characteristics of the materials at 10 kHz*

4.6.6.2. *Near-field attenuation for aluminum and steel*

Table 4.9 presents the wave impedances and the attenuations through reflection in near field at 10 kHz in the case of shielding placed at a distance of 10 cm from the source. In electrical near field, due to the high wave impedance, the attenuation through reflection is very high for the two materials. In magnetic near field, the wave impedance is low, barely two times higher than that of steel, and the mismatch is much less apparent. The field manages to penetrate the shielding well, and it is the attenuation through absorption which must ensure the efficiency of the shielding. The thickness of the shielding is, therefore, the determining factor.

Near field	Electrical	Magnetic
Wave impedance Z_s	$18 \times 10^{6}\ \Omega$	$7.9 \times 10^{-3}\ \Omega$
R_{dB} Steel (10 kHz)	182 dB	1.2 dB
R_{dB} Aluminum (10 kHz)	220 dB	32 dB

Table 4.9. *Attenuation through reflection in near field at 10 kHz*

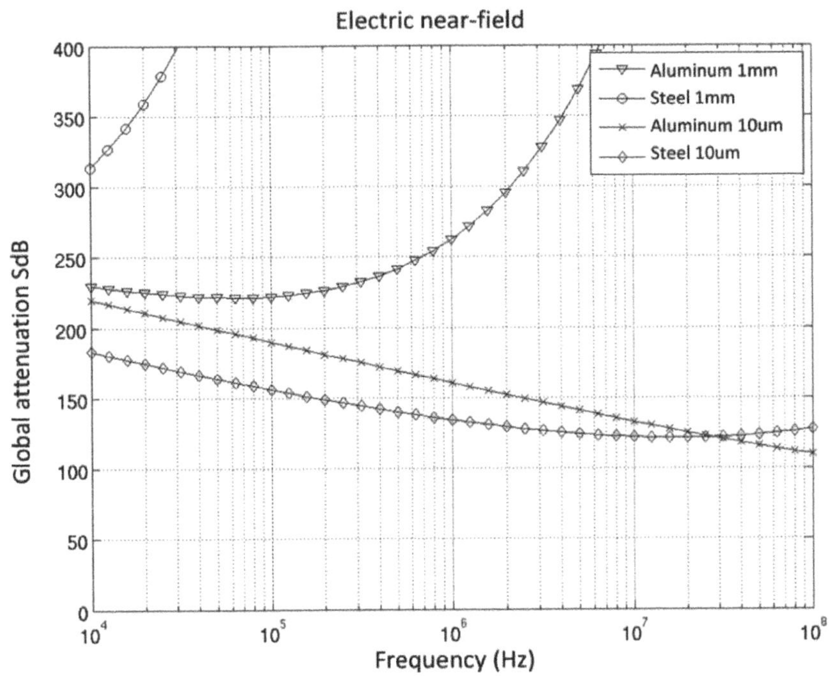

Figure 4.53. *Electrical near-field attenuation at a distance of 10 cm for aluminum and steel*

Figure 4.53 shows the evolution of the global attenuation in the case of electrical near field (source at 10 cm from the shielding). The attenuation caused by the reflection of the field is very high. Although this attenuation decreases with the frequency, the appropriate choice of thickness of the shielding lets us ensure a high attenuation over the whole frequency range without any difficulties. If we compare the behavior of steel and that of aluminum, we must note that steel causes less attenuation through reflection than aluminum, but it causes higher attenuation through absorption for a given thickness. Thus, Figure 4.54 compares the global attenuation of 1 mm aluminum shielding and 130 μm steel shielding, and we see that, from a frequency level of 500 kHz, the two manage to reach the same level of attenuation.

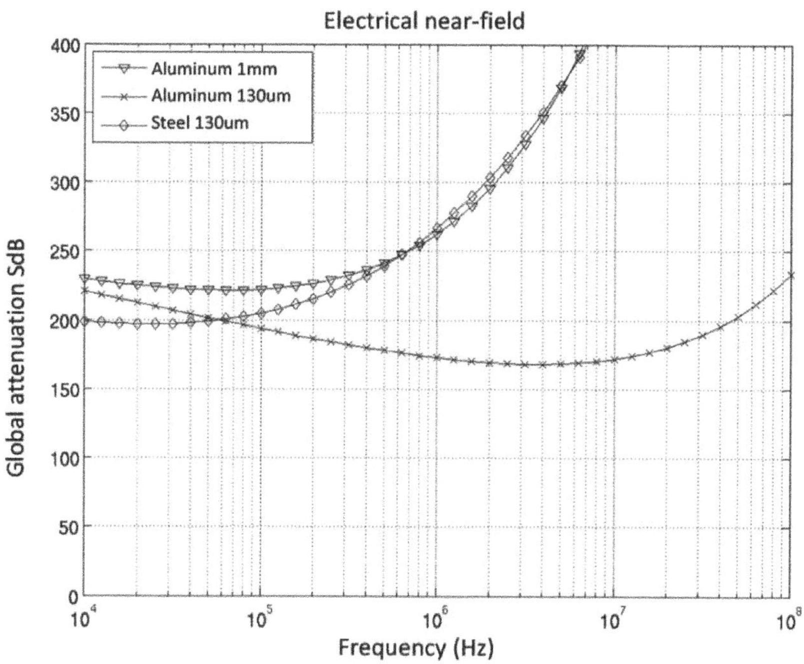

Figure 4.54. *Attenuation at 10 cm in electrical near field*

4.6.7. *Shielding issues for magnetic fields in low frequency*

Figure 4.55 shows the results in the case of magnetic near field and under the same conditions for the electrical field in Figure 4.53; the source is located 10 cm away from the shielding. Under these conditions, the wave impedance is $7.9 \times 10^{-3}\ \Omega$ at 10 kHz, low and very close to the wave impedance in steel (Tables 4.8 and 4.9), and there is practically no attenuation by reflection for steel. Even if the wave impedance in aluminum is higher, the attenuation by reflection remains limited.

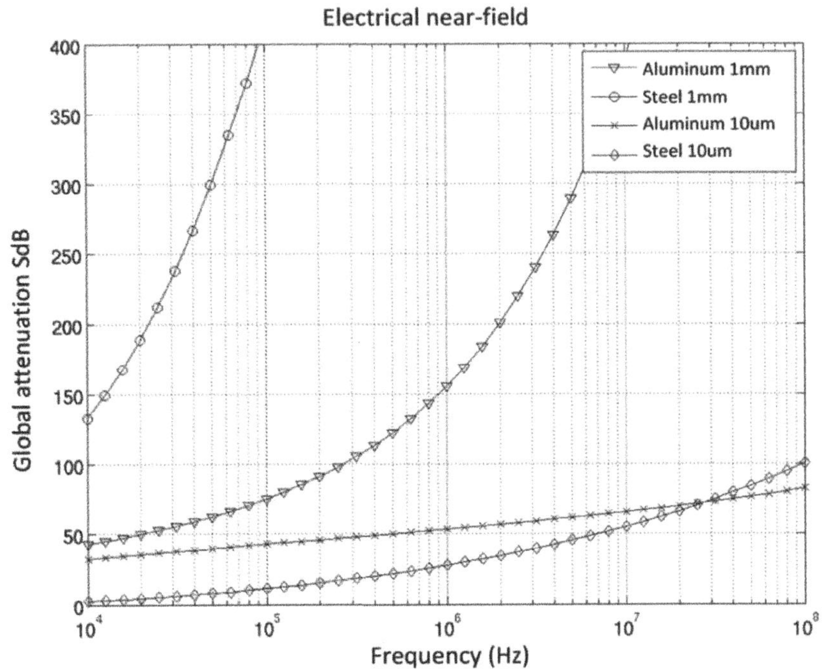

Figure 4.55. *Attenuation in magnetic near field at a distance of 10 cm for aluminum and steel*

The shielding of magnetic fields at low-frequency poses a real problem since for traditional materials, the effect of attenuation by reflection is low. Furthermore, the skin depth is increased at low frequency, and we therefore require high shielding thicknesses; special materials have been developed such as Permalloy and Mu-metal. They have a very high permeability at low frequency, which decreases rapidly with the frequency as shown in Figure 4.44. Figure 4.56 shows the results for Mu-metal compared with steel at low-frequencies (from 10 Hz to 10 kHz). Its high permeability helps to ensure a mismatch at very low frequencies and also has a higher attenuation by reflection than other materials.

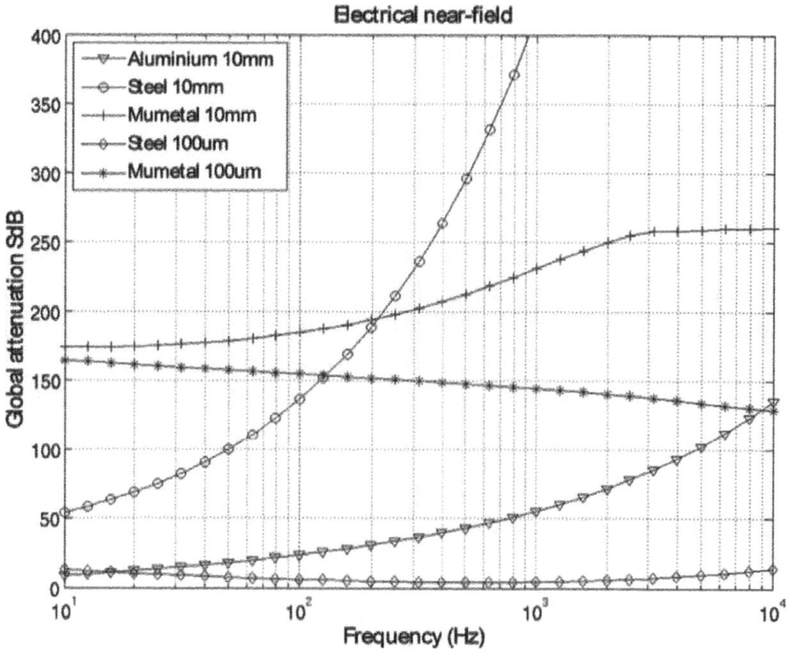

Figure 4.56. *Magnetic near-field attenuation (10 Hz–10 kHz) at a distance of 10 cm*

4.7. Conclusion

This section has enabled us to present the theoretical basis required for the understanding of physical effects that intervene in shielding. Due to the number of important parameters involved, it is not easy to initially determine the most efficient solution. Additional cost, volume or shielding weight constraints must be taken into account to choose a solution. Furthermore, the theoretical efficiency of shielding is limited by openings and the cables connected to the shielding [DEM 00]. These openings are required to ensure the ventilation of the equipment, but will limit efficiency at high frequencies, where the wavelengths will be of the same order of magnitude as the dimensions of the openings. We assume that an opening will begin to let frequencies pass, whose wavelength is less than 20 times the largest dimension of the opening (D):

$$\lambda_{limit} < 20 \times D \qquad [4.82]$$

For the through cables, we will, on the one hand, need to ensure the continuity of the shielding between the cable and the casing, with the best solution being a 360° ground connection, ensured using a specific connector, and on the other hand, ensure an efficient input filtering of the casing. We can find filtering EMC cards in certain types of equipment, which ensure this function for all connections between the casing and outside.

4.8. Bibliography

[CAN 04] CANTILLON-MURPHY P., NEUGEBAUER C., "An active ripple filtering technique for improving common-mode inductor performance", *IEEE Power Electronics Letters*, vol. 2, no. 2, pp. 45–50, June 2004.

[CEI 99] CEI, "Spécifications des méthodes et des appareils de mesure des perturbations radioélectriques et de l'immunité aux perturbations radioélectriques", CISPR 16-1 standard: 1999.

[COL 11] COLONEL MCLYMAN W.T., *Transformer and Inductor Design Handbook*, 4th ed., Dekker, New York, 2011.

[DEM 00] DEMOULIN B., DEGAUQUE P., "Blindages électromagnétiques", Archives D1320, Engineering Techniques, 2000.

[FER 99] FERRIEUX J.P., FORREST F., *Alimentations à découpage convertisseurs à résonance*, 3rd ed., Dunod, Paris, 1999.

[GRA 99] GRANDI G., KAZIMIERCZUK M., MASSARINI A., "Stray capacitance of single-layer solenoide air-core inductors", *IEEE Transactions on Industry Applications*, vol. 35, no. 5, pp. 1162–1168, September–October 1999.

[HEI 80] HEITZMANN J.P., "Cages de Faraday", Archives E1550, Engineering Techniques, 1980.

[LAB 10] LABROUSSE D., Amélioration des techniques d'estimation des perturbations conduites, application à une chaîne de traction de véhicule électrique, PhD Thesis, ENS Cachan, 2010.

[MAG 13] MAGNETIC MATERIALS DATASHEET, "Soft ferrites and accessories", *Data Handbook*, available at www.FerroxCube.com, 2013.

[MAS 97] MASSARINI A., KAZIMIERCZUK M., "Self-capacitance of inductors", *IEEE Transaction on Power Electronics*, vol. 12, no. 4, pp. 671–676, July 1997.

[SHE 04] SHEN W., WANG F., BOROYEVICH D., *et al.*, "Optimizing EMI filter design for motor drives considering filter component high-frequency characteristics and noise source impedance", *19th Annual IEEE Applied Power Electronics Conference and Exposition, APEC'04*, vol. 2, pp. 669–674, 2004.

[WAN 04] WANG S., LEE F.C., ODENDAAL W.G., "Effects of parasitic parameters on EMI filter performance", *IEEE Transaction on Power Electronics*, vol. 19, no. 3, pp. 869–877, May 2004.

[YOC 03] YO-CHAN S., SEUNG-KI S., "A new active common mode EMI filter for PWM inverter", *IEEE Transactions on Power Electronics*, vol. 18, no. 6, pp. 1309–1314, November 2003.

Index

F, H

I, L, M

N, O, P

Q, R

S, T

U, V, W

Printed and bound by CPI Group (UK) Ltd, Croydon, CR0 4YY

13/09/2026

14961075-0004